Innovation in Clusters

Smart Innovation Set

coordinated by
Dimitri Uzunidis

Volume 36

Innovation in Clusters

Science–Industry Relationships in the Face of Forced Advancement

Estelle Vallier

iSTE

WILEY

First published 2021 in Great Britain and the United States by ISTE Ltd and John Wiley & Sons, Inc.

ISTE Ltd
27-37 St George's Road
London SW19 4EU
UK

www.iste.co.uk

John Wiley & Sons, Inc.
111 River Street
Hoboken, NJ 07030
USA

www.wiley.com

Library of Congress Control Number: 2021943916

British Library Cataloguing-in-Publication Data
A CIP record for this book is available from the British Library
ISBN 978-1-78630-625-8

MIX
Paper from responsible sources
FSC
www.fsc.org
FSC® C013604

Contents

Foreword

It is a peculiarity of these pandemic times, but the word "cluster" has certainly never been in such widespread use as it is today, from epidemiological science to the most common conversations, describing the development of a virus called Covid-19. However, in other times and places, and with regard to other fields of human activity, this fetish word has been used to saturation point, even if today, in certain situations, it gives way to a more trendy word: ecosystem. We may as well admit that the more than extensive meaning of this term is quite appropriate, given the various projects commissioned by any number of organizations or specific groups. This is particularly the case in those localized instances where science, industry and public authorities are, in certain fields of technosciences, attempting to work together, as much by smart invocation as by physical methods of communication. Their ambition, resolutely symbolized and displayed on their letterheads or websites, is to attract, fertilize and develop innovation aimed at certain fields of science, which, ultimately in the form of goods, will be commercialized with the help of devices, procedures, advice and financial resources.

This is indeed the purpose of this rich and fascinating book by Estelle Vallier, structured in two parts and six chapters. Its title clearly invites us to explore the lineaments of this type of three-way collaboration dedicated to the production of innovations, to study its various mechanisms and movements, to understand its concrete pathways, to consider the institutional and economic goals it demands and, finally, to identify the obstacles that arise in practice for the parties involved in order to better understand them. In short, to shed new light on the processes at work which, to begin with, we would summarize under the title Technoscience Clusterization.

Certainly, some may argue that a vast literature already exists on this subject and that there is no need to attempt to further understand what has already been understood and, to a certain extent, the first two chapters of this book could bear witness to this. However, far from the sometimes condescending repetitions and unstimulating writing conventions, on the contrary, an astute state of the art is presented to the reader. Articulated through both historical and international perspectives, it offers from the outset a point of view and, consequently, an intelligibility so that this global phenomenon of technoscience clusterization can be understood, which, nevertheless, because it cannot be grasped on an international scale, is rooted in national and local institutional and organizational logic. It therefore enables soothing discourse to be avoided and, in contrast, an understanding, on the one hand, of how this accumulated knowledge, particularly in management sciences, has been able to accompany and justify its general design, from which some miracle had been expected, and, on the other hand, its concrete and localized implementation by *ad hoc* public policies. It is clear that the approach chosen by Estelle Vallier is congruent with the deployment of her research; for, rather than aiming to celebrate once again the success of this three-way marriage, her ambition was to renew questioning by seeking to interpret the difficulties that the actors in this complex relationship face in practice in order to collaborate. This is indeed the central issue that is anticipated from institutionalized science–industry relationships looking to produce any desired innovation.

This was the starting point of this research, described in the Introduction, for it was necessary for a cluster pilot study to question its own role, to succeed by depending somewhat on an ambient imagination (an ideology?), made to measure like any readymade garment, before being able to call upon the critical and interested gaze of social science. This book is specifically the end result of such sociological research, certainly commissioned, as well as negotiated. This research is rooted in France's best-known biocluster, which possesses the legitimacy that comes from its 20 years of existence: Genopole, based in Evry, in the south of Paris. In practice, the research posture of active participation in cluster coordination, while at the same time succeeding in maintaining a distance from it, could be more than embarrassing, given the risk of not being able to keep the two together over the long-term. Methodologically, this dual position has, in the end, enabled this collective role, the initial aim of which was the steering of the cluster, but entirely diverted towards the success of the relationship between science and industry, to be explored in context.

Therefore, we can say without further ado that the challenge has succeeded. This result, through the detailed analysis developed throughout the chapters of the second part of this book, leads the reader to explore the different levels of intermediation through which any institutionalized cluster pilot operates. As a result, the different problems and obstacles that resolutely stand in the way of the imagined and imaginary road of the uninterrupted flow of innovations from science to industry, in which it continues to believe, are revealed. Throughout the chapters, the reader will be aware of this need for creativeness on the part of the cluster's operational teams, in order to keep the flame of successive attempts to establish links alive. The picture that emerges is similar to the reproduction of the legend of Sisyphus. This is evidenced by the many meeting mechanisms, tools and instruments of all kinds, all of which are aimed equally and simultaneously at academic laboratories and enterprises (mainly start-ups) that have been labeled, rather than those that are in the process of being labeled. Moreover, but no less important, lies a paradox specific to this instituted and organized process; for if the stated aim, the "cluster effect", is to generate innovations from this networking, without this aim being able to be truly achieved to its desired extent, what hinders or even prevents it, without knowing how to precisely identify its contours, obliges the intermediary to demonstrate innovation himself in order to succeed in finding the right passage. However, it is one thing to note a certain impasse in this path and quite another to explain it. Perhaps at this nodal point, armed with the analysis, we see the specific and advantageous contribution of social science, singularly of sociology.

The final two chapters propose unravelling what is constantly being tied, rewoven and replayed by the practices of the actors themselves: scientists, startup entrepreneurs and cluster administrators. It is clear that what appears to them as an enigma pushes them to redo what has already been done, through the use of different devices. The path taken by Estelle Vallier to provide reason to this process, which is to an extent blocked in on itself, is to question cooperation, as a generic relational mode in a world given over to competition.

However, there is no better plot, including in social science, than to leave it to the reader to discover the mechanisms and the solutions that result from them. I will therefore refrain from going any further, since the reader of this book could be a sociologist, of course. Let us hope, however, that beyond the legitimate concern of satisfying the interest of subject matter specialists, practitioners of technoscience clustering will themselves be able to discover, from another perspective than their own, this intertwining that they contribute to produce, together, in their day-to-day relationships. For, if some consider that sociology is a possible weapon for transforming social practices, there are indeed some reasons. The simplest to state, but certainly, it seems to me, the most difficult to implement, is this: above all else, the role of sociology is to contribute to enriching the reflexive process of the action of any individual and/or collective. In this, it is fully involved in the social future and must claim it. There is no doubt that this book bears the mark of this!

Philippe BRUNET
Laboratoire interdisciplinaire
Sciences Innovations Sociétés
Gustave Eiffel University
August 2021

Introduction

Innovation is as much an overused term as it is an economic watchword: "We must innovate, be innovative… or disappear!"[1] It is a sign of economic recovery, dynamism and technological potential for a better world. One of the main levers of innovation policies in industrialized countries is to organize meetings between researchers and industrialists, particularly within innovation clusters that bring together companies, laboratories and training within a particular sector. Thus, the validity and legitimacy of these mechanisms, which take shape in dedicated locations, formerly called technopoles, now known as clusters, no longer need to be proven. Tomorrow's economy, like a promise that is always being renewed, is germinating in these places through the combined effects of innovation, its territorialization and the power of the intricate and informal relationships that are generated there. This book is particularly concerned with the articulation between the promise of innovation and the interactions observed within a particular cluster.

I.1. Innovation policies and the clustering process

In France, innovation clusters are the result of a legislative and fiscal framework that has encouraged closer ties between science and industry for more than 30 years. In this context, the state, as well as international and European authorities, is particularly encouraging territories to develop clustering policies. However, this logic is not a new idea and has been the subject of work since the 19th century, under other names. The term cluster,

1 Formula found on the back cover of Gaglio (2011).

on the other hand, was conceptualized in North American management literature in the 1990s and has found favor in most public policies in industrialized countries.

I.1.1. *Ensuring the legal and fiscal framework for the partnership between science and industry: governing from a distance*

In recent decades, most industrialized countries have opted for strategies that boost the potential outputs from innovators or creators (goods and services resulting from innovation) by legislating and investing in them. The main public investment consists of aligning the creation, or scientific discovery phase as closely as possible to its technological and industrial application. Efforts are essentially expected from researchers, who are encouraged to become entrepreneurs, or at least to make their research immediately available for industrial use (Stehr 2000). Thus, the expression "contextualization of science" (Gibbons *et al.* 1994) has been coined, within European institutions, to describe the process of heteronomization of science by other social fields, principally economic; it must be accountable to society and, above all, be involved in the commercialization of its results (Lamy and Shinn 2006, p. 47). In France, the 1999 Innovation and Research Act, also known as the "Allègre Act"[2], is presented as the significant legislative tool that guarantees a better contextualization of science. Indeed, it enables researchers, based on their scientific results, to create a company and/or to register patents. In order to support these new kinds of entrepreneurs in the creation of their start-ups, the Act provides for the reinforcement of the creation of public incubators. These structures are intended to provide, not direct financial, but indirect resources to assist in the setting up of a company: legal advice, construction of a business plan, filing of patents, search for public and private funding, etc. Incubators also provide privileged access to material resources, such as premises or equipment. This legislative framework complements the advantageous tax structure provided to research and development (R&D) since 1983 with research tax credits (*crédit d'impôt recherche*, CIR). Tax reduction, which is granted to companies based on their R&D expenditure, is the largest financial investment by the government in this area, especially since the 2008 reform

2 The Minister of Higher Education at the time.

under President Sarkozy[3]. In a 2013 report[4], the Court of Auditors highlighted the significant tax advantage represented by CIR, which primarily favors large companies[5], without any real impact on the recruitment of scientific personnel in the private sector, particularly young doctors. Despite numerous warnings and limitations, demonstrated on several occasions by research actors, including a report by the CNRS and a survey report by the researchers' association "Science en marche *(Science on the move)*" (Métivier 2015), and by institutional actors[6], it was renewed by the Valls government in 2015. This renewal demonstrates the political vision of innovation and its public governance, which, despite criticism from higher education circles, favors the commercialization of research within private companies.

Indeed, French innovation policies are characteristic of governmental adherence to neoliberal rationale. Often represented as a doctrine of state withdrawal, it does not, however, manifest itself in a paralysis of public authorities. On the contrary, their involvement is redeployed and takes a different form: paradoxically, "while it advocates the withdrawal of politics and the state, neoliberalism frequently results in their reaffirmation" (Epstein 2006, p. 96). Rather than guaranteeing the distribution of wealth among its territories and population, the state is actively providing a legal, social and fiscal framework to foster attractiveness, competitiveness, innovation, entrepreneurship, etc. (Bruno and Didier 2013, pp. 38–39). To do this, the state governs remotely (Epstein 2006) via high-tech Colbertism (Cohen 1992) in sectors deemed promising for the economy of the future: digital, biotechnology, nanotechnology, Big Data, etc. It is becoming an actor in global competition, through a selective location policy, which aims to attract the most innovative companies and the most affluent social classes to these territories, in order to stimulate a new development model (Brenner 2004).

---

3 Since 2008, the CIR has been calculated as follows: it is equal to 30% of research expenditure less than or equal to 100 million euros and equal to 5% of expenditure above 100 million euros.

4 Report to the Assemblée's nationale's Finance Commission, "L'évolution et les conditions de maîtrise du crédit d'impôt en faveur de la recherche", Cour des Comptes, July 2013.

5 Some companies, such as Intel and Sanofi, have been singled out in the press for laying off employees while benefiting from CIR.

6 The risks of fraud linked to the size of the system were pointed out in the report of the Court of Auditors and, in 2014, a Senate Commission of Inquiry voted not to publish a report on the reality of the deviation from CIR.

This is by no means specific to France. The scope of success stories such as Silicon Valley, and the legend around the innovators who have emerged from it, generally ignores the massive and historic public investment in the United States (Mazzucato 2015). Thus, the European Union, through two major projects, the Bologna Process and the Lisbon Strategy, intends to provide the framework for the development of a knowledge-based economy. Public policies in favor of the concentration of innovative activities on technical–scientific–industrial sites (Sainsaulieu and Saint-Martin 2017, p. 11), or, in a word, clusterization, represent one of its tools. France has not escaped this and is renewing its policies, as it did in the past with the nuclear and aeronautical industries, by relying on the concept of clustering in order to develop and position innovation (Lamy and Le Roux 2017, p. 89).

I.1.2. *Clustering: an old idea at the heart of current innovation policies*

High-tech clusters have been the subject of a wealth of research, principally in economics and management sciences, as well as in geography and sociology. Much of the literature can be traced back to the neoclassical economist Alfred Marshall and his concept of the industrial district, considered to be the ancestor of the cluster. As early as 1890, he defined it as coordination by the market and the reciprocity of a social division of labor between small specialized firms within a large productive process (Benko *et al.* 1996, p. 120). His approach was revisited in the 1970s by Italian[7] sociologists and economists as a framework for analyzing their research on the Third Italy[8], which specialized in traditional (clothing, footwear, leather, furniture, etc.) or more modern (small-scale mechanics, electrical engineering, etc.) activities (Daumas 2007). Based on Marshall's work and his analysis of the province of Prato in Tuscany, Beccatini defines the Italian district form as follows:

> A socio-territorial entity characterized by the active association, in a circumscribed and historically determined territorial area, of a community of people and a population of industrial enterprises. In the district, unlike what happens in other

7 Arnaldo Bagnasco, Carlo Trigilia, Sebastiano Brusco and Giacommo Beccatini.

8 Expression referring to the regions of Florence, Bologna, Venice and Treviso, between the prosperous and industrial Italy of the North and the poorer Italy of the South.

> environments, such as the manufacturing city, community and business tend to, as it were, interpenetrate (Becattini 1989, p. 15).

This idea of interpenetration between economic structures and individuals is taken up again in policies of technopolization and then clusterization. While the work on districts was aimed at analyzing existing environments, technopoles are the first political attempts to develop niches of specialization, in order to foster social and economic environments conducive to exchanges. Moreover, it is no longer just a matter of prioritizing inter-firm relationships in the Italian districts, but of reproducing the American science–industry relationship models. The most analyzed, of these, during the 1990s, were Silicon Valley in the San Francisco Bay Area (Saxenian 1996; Ferrary and Pesqueux 2004) and Route 128 in Boston. In these studies, the emphasis is on the concentration of companies in the same sector, as well as on the geographical proximity of research laboratories and educational institutions, so that informal relationships emerge that promote innovation, collaboration and the creation of businesses. In 1990, in his book *Competitive Advantage of Nations*, Michael Porter placed the local concept of the cluster at the center of his theory of competitiveness within the context of a globalized economy (Porter 1990). His definition is internationally understood by public decision-makers in industrialized countries:

> Clusters are geographic concentrations of interconnected companies, specialized suppliers, service providers, firms in related industries, and associated institutions (for example, universities, standards agencies, and trade associations) in particular fields that compete, but also cooperate (Porter 1998b, p. 4).

Porter's conceptualization is seen in some geographic literature as a belated rediscovery of well-established evidence (Torre 2006, p. 17), including analyses of districts or technopoles. Nevertheless, this updated definition has been taken into account by public policy to the point of becoming a normative concept. It satisfies the ambition of reindustrialization, based on scientific innovation and a flexible network of small and medium-sized enterprises (SMEs) adapted to the global economic system (Benko and Lipietz 1992), notably in Germany (skills clusters), Japan (knowledge clusters) (Forest and Hamdouch 2009) or even in the United Kingdom, where it inspires New Labour in its objective of renewing industrial policies (Zepf 2011, p. 187). In France, the cluster concept also emerged at the end of the 1990s under a socialist government[9]. Some aspects of the concept are alluded to through, as seen above, the creation of public incubators in the Allègre Act of 1999. In 2004, the concept was firmly established at the heart of the competitiveness cluster system. A characteristic of remote government, the state intervenes by calling for projects, designed to grant subsidies to selected candidates representing consortiums of companies, laboratories, organizations, associations, etc., from the same sector of activity. The call for projects is both a way to empower local actors, while also reducing their autonomy, insofar as the competing territories must rigorously respect the specifications established by the state if they wish to be selected (Epstein 2006, p. 108). In 2004, of the 105 applications submitted, the state selected 66 (there were 67 in March 2018)[10]. Applicants are well aware of the U.S. references linked to the competitiveness cluster system, and this is how the *Valley* projects appear: Aerospace Valley, Alsace Biovalley and even Cosmetic Valley.

Once they have been put in competition with each other in a general call for projects, the territories develop strategies to find the "right ingredients"[11] that promote innovation at the cluster level. The three essential dimensions of the typical ideal cluster are industry, science and education. They are embodied by companies, usually a network of SMEs and very small enterprises (VSEs), but the presence of large groups is often desirable for the cluster's reputation and attractiveness, by both private and public research laboratories, and by renowned higher education institutions[12]. Among the

---

9 On this point, see the article by Étienne Vergès on the links between the socialist governmental majority and research valorization policies, from the Act of July 15, 1982, to the Fioraso Act of 2013: Vergès E. (2014). Normes de la recherche scientifique, *Cahiers Droit, Sciences et Technologies* [Online]. Available at: http://cdst.revues.org/346 [Accessed September 30, 2016].

10 The 2005 Finance Act defines competitiveness clusters as "groups of companies, higher education institutions and public or private research organizations within the same territory, committed to work in synergy in order to implement economic development projects for innovation" (Finance Act No. 2004-1484 of December 30, 2004).

11 This metaphor of a recipe to be reproduced by combining the right ingredients is present in the political discourse around clusters, as well as in that of the business leaders and laboratory directors who operate in these spaces.

12 To these three basic ingredients could be added, depending on the space, companies providing support functions (legal, IT, etc.), associations, hospitals (in the case of health clusters), administrations or public bodies, etc.

definitions of clusters found in the literature, and without claiming to be exhaustive, the definition given by Lamy and Le Roux in 2017 appears to be one of the most complete:

> It is therefore a question of a favorable environment that is not limited to elementary communication infrastructures but which benefits from the immediate proximity of specialized actors and expertise. These benefits are expected in the form of a ripple effect (called a "network" effect by economists) by mutual reinforcement of several factors: productivity gains (through proximity, mutualization and/or reduction of intermediaries); the pooling of skills (under the auspices of the availability of human resources, but also of the maintenance of a permanent motivation based on challenge and elitism); an intensification of the circulation of information (which is important for prospecting and coordinating markets, but also for sharing tacit knowledge), etc. The aim is to obtain chain reactions of innovations, each one reverberating in this environment and acting as a sounding board to encourage others. The attractiveness of the cluster and its self-amplifying character must reach a critical mass providing a sustainable global competitive advantage (increasing visibility and therefore attractiveness, etc.) (Lamy and Le Roux 2017, p. 91).

Here, we see the concept of an environment that is "favorable" to innovation, where geographical proximity is a key factor in the interpenetration of the various neighboring structures. This intricacy enables the sharing of equipment and the circulation of individuals between structures. Underpinned by implicit knowledge sharing, it leads to innovations, which are formal and wealth-producing and which reinforce the attractiveness of the cluster.

I.1.3. *Focusing on biotechnologies: catching up with the world through clustering*

This book is particularly concerned with the contextualization and clustering of the life sciences. A major turning point for the life sciences occurred in the 1980s in the United States with the passage of the Bayh-Dole

Act, which allowed universities to patent publicly funded research (Gaudilliere and Joly 2006). The patenting of life forms has thus played an important role in bringing together the life sciences and their technological application, commonly known as biotechnology. This is understood as "a set of procedures implementing biological knowledge for a transformation of the living body: gene therapy, cloning, genetic modification of organisms, etc." (Keck 2003, p. 182). It is identified as a field with a high technological potential, capable of "irrigating large areas of the economy" (Branciard 2004, p. 10), its aim being to make living organisms a factor of production like any other. Since 1983, the Ministry of Research has notably been one of the institutions involved in promoting biotechnology through organizations such as the National Scientific Research Centre (*Centre national de la recherche scientifique*, CNRS), the Atomic Energy Commission (*Commissariat à l'énergie atomique*, CEA), the National Agronomic Research Institute (*Institut national de la recherche agronomique*, INRA), the National Health and Medical Research Institute (*Institut national de la santé et de la recherche médicale*, INSERM), universities, foundations, etc. In addition to the creation of a national committee on these issues (the "biotechnology boom" program)[13], the state committed itself to promoting biotechnology on October 17, 1990. Hubert Curien, Minister of Research at the time, launched the National Human Genome Program in the Council of Ministers (Guthleben and Faou 2011). The aim was to set up a coordinated strategy with all the laboratories and institutes already involved in biological research on the genome. The program also aims to coordinate cooperation with the United States around the Human Genome Project, which aims to establish complete DNA sequencing.

Biotechnology generally refers to health-related production processes, but in reality, it covers other industrial sectors, primarily biopharmaceuticals (therapeutic products, diagnostics, medical devices, etc.) and the biofood sector (Heil 2010, p. 240). The former comprises two types of actors: small innovative firms often originating from academia (spin-offs) which, in most cases, once the fundamental research has been completed, are bought out by "big pharma", which takes over the costly processes of the clinical phase,

13 https://www.siv.archives-nationales.culture.gouv.fr/siv/rechercheconsultation/consultation/ir/pdfIR.action?irId=FRAN_IR_019764.

production and marketing. The latter addresses human nutrition, agriculture and the environment (production of food based on microalgae, detection of plant pathogens, biorefining, detection of pollutants from modified living organisms, etc.).

Biotechnology assumes both a strong academic base and interaction with the medical and industrial worlds (pharmaceuticals, therapies, agrifood, agrochemicals, environment, bioenergy). It is therefore positioned between a world that guarantees diversity (academic research) and another whose challenges, conversely, relate to standardization (medical and/or industrial application) (Branciard 1999, p. 3).

In order to resolve this conflict between standardization and the maintenance of diversification, public action mechanisms create conditions for the coming together of laboratories, universities, companies and equipment in localized spaces (biotechnology clusters or bioclusters). The cost of equipment, in particular, would explain the need for agglomerated networks in the case of biotechnology, in order to pool vital, cutting-edge instruments (Aggeri *et al.* 2007b, p. 202). Finally, the dominant argument in the 1990s was that there was too great a gap between scientific production in the life sciences and its commercialization, which was reflected in an insufficient number of start-ups and patent applications. Particularly in the case of biotechnology, the rhetoric is as follows: "The articulation of science with industry is not self-evident: a forcing seems necessary" (Brunet 2011, p. 2). From then on, bringing together two worlds that ignore each other requires incentives and, in particular, the creation of an intermediation structure between science and industry.

I.2. The cooperation mechanism in a biocluster context: from concept to reality

Originating in the United States, the cluster concept has become a worldwide phenomenon. In France, it has been supported by public authorities through the creation of intermediation structures in charge of strengthening the relationship between science and industry.

I.2.1. *The advent of structures for science and industry intermediation*

The literature uses the term "hybrid organization" (Branciard 2009) or "hybrid organism" (Leydesdorff and Etzkowitz 2000)[14]. Very quickly, within institutional vocabulary, the term "incubator" has become a cornerstone of science–industry rapprochement policies (Shinn 2002, p. 28). Nevertheless, among these different terms and, in a sociological approach, that of "public intermediation structure", Brunet pertinently evokes the intervention of these structures between ("inter") two quite distinct worlds (Brunet 2011), in order to facilitate arbitration intended to reconcile them ("mediation")[15]. Moreover, this name highlights the public character of these institutions.

These structures, which are generally financed by local and national governments, can be found in various forms (associations, mixed syndicates, mixed economy companies, etc.). Most were created in response to calls for projects from competitiveness clusters. However, some were created before, and in parallel with, these clusters. The latter are anchored in a regional framework, but there are other structures for which territoriality is defined on a conurbation or a department scale. Most operate with an accreditation policy to attract laboratories and companies. Once accreditation is obtained, the new member organization benefits from services offered by the cluster (preferential rents, access to specialized and shared equipment, a computer network, etc.). When local geographic clustering is carried out, most structures rely on the "total network" (Suire and Vicente 2014). The relational density between the three actors of science, industry and training, enabled by geographical proximity, is touted as a major condition for innovation. Nevertheless, many intermediation structures struggle to meet this objective and find themselves confronted with the following organizational difficulty: how to create cooperation when it does not occur spontaneously through geographical proximity. This question echoes sociological literature, which highlights the absence of spontaneous links between spatial proximity and social interaction, notably in the study of the

14 In their triple helix concept (three helices representing academia, industry and government), Leydesdorff and Etzkowitz consider that the central space where the helices overlap forms hybrid organisms that are intended to make the articulation of the three spheres more effective. This concept will be further detailed in Chapter 1.

15 http://www.cnrtl.fr/etymologie/interm%C3%A9diaire.

politics of large urban areas (Chamboredon and Lemaire 1970). Work on clusters also emphasizes the relationship or otherwise between, for instance, geographic, organized and cognitive proximity (Talbot and Kirat 2005; Bouba-Olga and Grossetti 2010; Torre and Zuindeau 2012).

This was the problem encountered by Genopole, a life sciences biocluster located in Evry, in the Essonne region of France, the main field of investigation for this book, which is the result of a thesis in sociology funded by Industrial Agreements for Training through Research (*Convention industrielle de formation par la recherche*, CIFRE) and which reports on the results of an immersion survey conducted over three and a half years, between November 2013 and April 2017. Genopole is a cluster that has already been studied in the literature, particularly on the political, economic and social conditions of its creation and institutionalization, mainly in the work of Anne Branciard (1999a, 1999b, 2002, 2004, 2009), as well as of Ashveen Peerbaye, which sheds light on the instrumental arrangements put in place by the cluster (Peerbaye 2004). Genopole has also been the subject of a comparative analysis of the transformation of science into technoscience, based on the cases of Evry, Laval (Quebec) and San Diego (California) (Heil 2010). There is only one other study that deals with its interactional dimension in a comparative approach. The article in question focuses on the "social capital of entrepreneurs as an index of cluster emergence" in a "comparative analysis of the transformation of two bioparks into bioclusters: Kobe (Kansai, Japan) and Evry (Paris region, France)" (Lanciano-Morandat *et al.* 2009). In their conclusion, the authors state that, although they have their own national characteristics, neither the Kobe nor the Evry bioparks can be categorized as innovation clusters as defined in literature insofar as:

> They are still only aggregates of diverse entities with episodic relationships between them; in short, partial clusters [...]. Both parks are struggling to integrate the entire innovation process, which, as Porter suggests, is a condition for their constitution as clusters. In addition, both parks have the weight of the state in their constitutions in common, the role of certain institutions in their creation, and their location on sites that have neither a tradition nor particular resources in terms of innovation (Lanciano-Morandat *et al.* 2009, p. 200).

In addition to the fact that the authors rely on the definition proposed by Porter to identify, or not, the clustering process, the idea of transformation from biopark (geographical grouping) to biocluster (interactions) is also strongly present in management literature (Hamdouch 2007). These concerns are also at the heart of current biocluster issues. Indeed, considerable space is given to this subject in formal and informal discussions within the cluster, warning about the difficulties of mobilizing companies and laboratories for joint meetings.

I.2.2. *From the cluster concept to its realization: between adoption and resistance*

This book therefore rightly proposes to observe and report on the application of the cluster concept within a specific field. The aim is to revisit the construction of this public action mechanism, which seems to be unanimously accepted, at least in discourse on innovation policies, and to confront it with the dynamics of cooperation in one of these clusters. The purpose of the book is thus reminiscent of the study of the "editing work" of foreign examples (Sahlin and Wedlin 2008). This work consists, particularly in the field of science policy, of adapting a public action mechanism already implemented abroad to the context of another country, region, city, etc. The authors speak of operations of decontextualization and recontextualization of international policies in order to inscribe them into another national or local framework. The work of editing allows us to focus on the actors who construct and "edit" these foreign examples, and on the importance of the latter in the adoption of funding policies, in particular. Séverine Louvel and Mathieu Hubert have particularly shown the role of foreign examples in the implementation in France of nanoscience and nanotechnology steering (Louvel and Hubert 2016). In the same way, this book proposes to revisit the construction of the cluster concept on an international scale and to find out who the main authors and disseminators are. However, it also proposes to report on the reality experienced by those who work within these geographical clusters.

To do this, the second part of the book draws more on the literature on the sociology of work than the first part, which is mainly based on the sociology of science; effectively, how to impose a cooperation system on individuals who work for different employers (public laboratories, start-ups, SMEs, private laboratories, associations, etc.) and who, what is more, belong to two different social fields: scientific and economic. The field survey shows that these individual affiliations are often in contradiction with the cluster's objective of promoting synergies. Recourse to the work of sociologists who have already highlighted this type of paradoxical injunction specific to modern management (Linhart 2010, 2015; de Gaulejac and Hanique 2015) or disembodied management (Dujarier 2015) has made it possible to study the effects of the networking system on the experience and practices of individuals in the context of the cluster.

I.2.3. *An immersion survey: observing, interviewing and quantifying on a daily basis*

This survey, which was carried out as part of a CIFRE sociology thesis, is based on a variety of empirical material and combines qualitative (semi-directive interviews, participant observation, documentary and archival database of the biocluster) and quantitative (questionnaires and network analysis) methods.

On the qualitative side, the survey is based on 45 semi-structured interviews with a large proportion of laboratory directors and company managers, as well as with scientific directors, operations managers, chief financial officer (CFOs), communication and/or marketing managers, platform managers, research engineers, doctoral students, management assistants, physicians, etc.

These interviews were supplemented by a long phase of participant observation as a doctoral student at Genopole. This immersion was both conducive to the observation of official speeches (during colloquia organized by Genopolemeetings, visits by elected officials, etc.) and to participation in informal discussions. The confrontation of these two types of material gave rhythm to the research as a dialectic between political discourse and the reality of practice. In addition to observation, being on site provided access to certain documents archived by Genopole (activity reports since 1998, meeting minutes, press releases, digitized official documents, etc.) and to its internal database (management charts, files of accredited members, working documents, etc.).

In parallel, more quantitative material was collected over various stages of the survey. The first was a questionnaire, completed by the Genopole teams, on the expectations of employees in terms of coordination between the site's structures, completed by 534 people. Although the sociological scope of the survey has its limits (due to its operational purpose), the number of respondents allows for a significant increase in generality.

Nevertheless, the quantitative material relies primarily on the network analysis method, which enables social interactions to be formalized using nodes and links on a graph. Nodes typically represent individuals or institutions, with links representing a particular type of interaction between two nodes. This method was adopted following the completion of a significant amount of qualitative work. As additional information became less and less significant with each new interview (saturation effect), it became essential, not to find missing data, but to observe the state of interactions at the cluster level. A relational database was therefore created on the basis of the ethnographic work carried out beforehand, and completed using the results of two questionnaires. The first, intended to observe the network at the inter-organizational level, was sent to all Genopole company and laboratory directors, the second to almost all the cluster's employees via the Genopole intranet site. The first survey provided relational data for 42 of the site's organizations[16]. The second survey collected responses from 102 people. However, the anonymity of the questionnaire did not enable the network to be mapped, but we were able to construct additional statistics based on the characteristics of the individuals interviewed and their relationship with the cluster[17].

Method	Material	Period
Observations	Research Officer attached to the General Management of Genopole	November 2013 to April 2017
Consultation on expectations in terms of coordination	534 responses from cluster workers	February to May 2014
Documentary and archival data	Genopole archived documents and internal database	May 2014 to April 2017
Semi-structured interviews	45 interviews	May 2014 to April 2017
Individual questionnaire for network analysis	102 responses from cluster workers	February to April 2016
Organizational Questionnaire for network analysis	32 responses from laboratory and company directors in the cluster	February to April 2016

Table I.1. *Summary of data collected*

16 The sampling method and data collection are detailed in Chapter 6.

17 Age, gender, profession, CSP, seniority, reasons for working on the site, the different sociability places or events and the type of relationships they have within the cluster.

Since clusters cover multi-scale (organizational and individual), as well as historical and institutional dimensions, methodological diversity appears to be essential in order to understand a plurality of sociological mechanisms. Taking only a single approach does not make it possible to determine a set of dynamics, as the authors of the article focusing on the social capital of entrepreneurs (Lanciano-Morandat *et al.* 2009) point out in their conclusion. Indeed, the articulation of empirical methods (monographs, interviews, network modeling, archives, etc.) allows us to move away from current representations of clusters and to better support certain hypotheses (Forest and Hamdouch 2009, p. 17).

Within the context of this book, the methodology used allows for a better analysis of the hazards of day-to-day cooperation. Indeed, the structure of the book is based on the tension, mentioned above, between the conceptualization and concretization of the cluster. The first part reviews the socio-historical dimension of the cluster and the contemporary diffusion of its precepts on an international scale, in the French context and in a local context in Île-de-France. The second part questions the operability of the model within a particular biocluster by analyzing the resistance to cooperation in the various laboratories and companies.

I.3. Acknowledgements

I would like to thank my colleagues at the Centre Pierre Naville: Émilie Balteau, Philippe Brunet, Fabrice Colomb and Nial Tekin, for their advice and proofreading, so that the original thesis could be transformed into a book. I would also like to thank the teams at Genopole for opening their doors to me and facilitating the investigation. Finally, thanks to Blandine Laperche and Dimitri Uzunidis of the Innovation Research Network for their reconnaissance and support in the publication of this book.

PART 1

Persistence and Renewal of the Cluster Concept in Contemporary Innovation Policies

1

From Industrial Districts to Knowledge Valleys: the Legacy of the Cluster

Innovation clusters are often presented as a new concept marking a break with a past, where companies and laboratories were not grouped together geographically and encouraged to interact. However, when we take a closer look at the history of spatial concentrations of industrial and technological activities, we realize that their contemporary character is not limited to the 21st century. On the contrary, the first examples date back as far as the 19th century. Whether we are talking about purely theoretical concepts or existing configurations, two types of cluster ancestors can be identified: the purely industrial concentrations, that can still be found today, and those that attempt, as early as the middle of the 20th century, to bring science and industry together within technological clusters.

1.1. The industrial district: the oldest ancestor of the cluster

Contemporary literature on innovation clusters often mentions that they were inspired by the industrial district, a concept dating back to the 19th century, the most significant examples of which emerged in 20th-century Italy.

1.1.1. *The economic approach of industrial atmosphere*

It is interesting to trace the history of the cluster back to the British economist Alfred Marshall, one of the fathers of the neoclassical economics. He developed the concept of the industrial district in one of his reference works, *Principle of Economics*, published in 1890. According to Marshall, there are two potential industrial organizations:

> On the one hand, the organization under the sole direction of the technical department integrated within a large enterprise. On the other hand, market and face-to-face (reciprocal) coordination of a disintegrated social division of work among smaller enterprises specialized in large segments of the production process (Benko *et al.* 1996, p. 120).

He then introduced a new element into the dominant thinking of the time, according to which productivity was linked to the division of labor within the enterprise (Marshall 2010, p. 119), asserting that work could be divided over a geographical area, between different entities. This is the industrial district, which he defines as the concentration of businesses and skilled craftsmen within a geographically limited space. Marshall sees three fundamental aspects to this: specialist employees, complementary industries and a permanent exchange of information and knowledge. These elements create a pool of skilled workers, thus making the district an efficient labor market. He explains the sustainability of the location of businesses in the same area using the principle of externalities and particularly economies of scale that would reduce production costs. Marshall distinguishes between internal economies of scale, derived from the resources of the businesses themselves, and external economies, obtained through the close proximity of similar but specialized businesses at a specific point in the production process. Moreover, since not all firms can afford all machines, the pooling of equipment allows each to increase its efficiency, without having to make heavy investments. Marshall's original, but insufficiently supported by empirical data, contribution is his concept of "industrial atmosphere", in which he describes the significance of knowledge transfer between enterprises, as well as transmission between colleagues, neighbors, members of the same family, etc.:

> The secrets of industry cease to be secrets; they are, so to speak, in the air, and children learn many of them unconsciously. Work well done is immediately recognized and the merits of inventions and improvements in machinery, processes, and general organization are discussed within the industry straight away. If someone comes up with a new idea, it is immediately

> taken up by others and combined with suggestions of their own and thus becomes the source of other new ideas (Marshall 1990, p. 160).

This therefore gave pride of place to interindividual and intergenerational relations, fostering an atmosphere conducive to learning and the dissemination of knowledge. Nevertheless, at the end of the 19th century, Marshall's industrial district was largely confined to a theoretical conceptualization on the eve of the Fordist period, based on the model of the large enterprise. It was not until the 1970s that the concept was revisited by Italian studies of the Third Italy, then taken up again in France.

1.1.2. *The first Italian districts and their influence in France*

At the end of the 1970s, sociologists Arnaldo Bagnasco and Carlo Trigilia (1977, 1984, 1988) and economist Sebastiano Brusco (1982, 1986) published their research on the Third Italy, the term used to describe regions referred to as intermediate because of their geographical location between the highly industrial regions of northwestern Italy and the Mezzogiorno, characterized by a lower level of economic development. The Third Italy specializes "in traditional (clothing, footwear, leather, furniture, etc.) or more modern (small mechanics, electrical engineering, etc.) activities, whose production process can be easily broken down" (Daumas 2007, p. 133). The industrial organization of this area is characterized by a strong presence of small and medium-sized enterprises that participate only in certain phases of production and do not integrate a vertical model into their processes (as opposed to the mass production processes of large factories, especially in automobile production, further north). This would therefore be a dynamic specific to small businesses (Bagnasco 1977).

These early studies emphasized the social characteristics of this region and, in particular, the family ties in the cooperation of businesses and the learning of trades. In these regions, described as "urbanized countryside", there would be a solidarity of the rural family at the time of the creation of an artisanal or commercial activity, succeeding or completing the original agricultural activity and then finding the political, logistical and social resources from medium-sized towns of the region. The network of parishes, especially in Veneto, is also highlighted as a source of information and exchange of know-how necessary for the creation of small businesses. The authors thus mobilize the Bourdieusian concept of social capital, as a "durable network of interknowledge and interrecognition relations of varying degrees of institutionalization" (Bourdieu 1980, p. 2), made available to potential entrepreneurs belonging to this local community. The focus is on interindividual relations and the cultural environment. Based on this work and his own analysis of the province of Prato in Tuscany, Giacommo Beccatini proposes a definition of the Italian district (Becattini 1989):

> A socioterritorial entity, characterized by the active association, in a circumscribed and historically determined territorial area, of a community of people and a population of industrial enterprises. In the district, unlike what happens in other environments, such as the manufacturing city, community and business tend to, as it were, interpenetrate (Daumas 2007, p. 134).

Studies on the Third Italy thus revive and further develop Marshall's concept of the district. They particularly emphasize the local accumulation of know-how, individual interdependencies and the interpenetration of businesses with a community of people.

Inspired by the Italian studies, research is also being conducted in France with the aim of identifying "territorial configurations similar to districts" (Guillaume 2008, p. 298) in other countries. This has led to work on "localized industrial systems" (Raveyre and Saglio 1984), as well as, under the impetus of a team from Grenoble, the economists Courlet and Pecqueur addressed "local productive systems" (Courlet and Pecqueur 1991, 1992). Raveyre and Saglio worked on the Oyonnax agglomeration, where industry was based on a single activity: the production of plastics. For them, this fabric of SMEs in the same sector cannot be analyzed as a scaled-down large enterprise, because the conflicts of power and the relations of competition and cooperation are not regulated in the same way as in large groups. These SMEs are involved in complex systems of relations (Raveyre and Saglio 1984, p. 159). For them, what is important is the local social act into which industrial firms are inserted, because defining a localized industry "is not a simple statistical aggregation operation" (*ibid.*, p. 160). They seek to look beyond the simple observation of a competitive situation and to determine the social construct of the localized industrial system, which is composed of a set of historically constituted standards and rules of conduct (*ibid.*, p. 168).

In the localized industrial system, collective mutual aid exists between firms in the event of difficulties or necessity.

On the other hand, Raveyre and Saglio consider that cooperation is not only defensive but also oriented towards strategies of development and adaptation of the technical potential of local industries (*ibid.*, p. 165). The system produces, through innovations, a disruption of the organization of work that translates into new trades and new training. Finally, another key element for the authors lies in the relations that the system maintains with local political authorities, which constitute a fundamental issue for the system's sustainability (*ibid.*, p. 174). In the local production systems studied by Courlet and Pecqueur (1992), the definition is broader than for the districts. They may therefore refer to similar activities in which enterprises cooperate on peripheral dimensions (transport, exports, etc.) or to complementary activities (Courlet 2002, p. 89) and, as a result, a network of subcontractors emerges (Guillaume 2008, p. 298). In some cases, businesses do not necessarily belong to the same sector and, unlike the district, do not participate in the multiple stages of industrial production. Moreover, the relationships are not familial, but professional and informal in a local productive system. In his work on the "Alpine furrow", however, the geographer Bernard Pecqueur (1995) qualifies this aspect and identifies relatively poor direct relations between businesses. Each has its own regional, national and even European and global geography, and what link businesses together are the territorial resources they share. Housing, amenities, infrastructure, facilities and the skilled labor market thus tend to unify them.

Despite some variations, both the Italian and French studies agree that the unit of analysis is not the enterprise, but the "district" as a whole, with interconnected firms (Brusco 1990). The focus has thus shifted from the enterprise itself to the space around it.

1.1.3. *The rise of districts: the end of the Fordist enterprise?*

Some authors see the sign of a more general economic change in the organization of industrial districts, in which Fordist mass production gives way to the industrial development of flexible specialization. This observation is mainly defended by regulationist economists (Boyer, Aglietta, Saillard, Coriat, etc.), who consider the accumulation regime concept as a tendency inherent to the capitalist mode of production, and which corresponds to regularities ensuring a general and relatively coherent progression of capital accumulation. These regularities operate within institutions that govern, among other things, competition, money, wage relations, etc. Depending on the place and the period, these institutional configurations may change. The Fordist mode of regulation articulated macroeconomic regularities on an international scale, while at the same time being subject to national compromises (Boyer and Saillard 2002). During the intense period of Fordism, the national scale was considered the relevant level of accumulation.

However, this economic and social regulation, which was being carried out within large enterprises, is no longer appropriate. Indeed, the crisis of Fordism is reflected in a crisis of territory at the level of the nation state (caused by a triple movement of tertiarization, deregulation and globalization) and a crisis of legitimacy of the welfare state with the emergence of mass unemployment (Carré and Levratto 2011, p. 360). In this context of capital internationalization and the crisis of the state, institutional arrangements are regional (EU, NAFTA) and international (WTO, for example), rather than solely national. In France, some authors see a disappearance of the Colbertist state (Mustar and Laredo 2002) and the emergence of new public actors, such as Europe and local authorities. As a result, the new post-Fordist regulations would be posed jointly in terms of sector and territory (Laurent and Du Tertre 2008). This approach is in line with theories that consider that a new geography of flexible accumulation is emerging in reaction to the Fordist mode of accumulation (Storper and Scott 1990). In a seminal work, Michael Piore and Charles Sabel (1986) develop the concept of flexible specialization as an alternative model to Fordist mass production, based on the emergence of local industrial models, notably the Italian districts (Hirst and Zeitlin 1992). Industrialization with flexible specialization is understood to be:

> A form of production based on individual initiatives or on those of small enterprises that will work together to build, through rather complex and informal transactional systems, products intended for the market (consumer products or production components) […]. This specialization in a segment favors flexibility in the production of the components and elements of the final product, which in turn leads to greater flexibility in the

> overall network of enterprises, which is then likely to offer a wide range of products to the market (Durand 1991, p. 48).

These SMEs, each specialized in a different segment of production, constitute a system that adapts to technological developments. For some authors, this specialization comes at a time of crisis for the mass consumption model: the rise of individualism leads the consumer to no longer want uniform goods (Capecchi *et al.* 1987). The weakening of these standardized goods markets has disrupted the large Fordist enterprises, the symbol of which being the automobile industry. Slow growth marked by weak demand would have determined the "shift of centrality" from large industry to small production units considered more flexible and more responsive to demand surges (Vercellone 2003). On the other hand, others consider that it is more a question of an entrepreneurial thrust that sought to express itself outside the conditions of domination of big business. Unable to compete on the basis of production costs, new production models were invented, based on product diversification (Durand 1991). Some observers agree that this post-Fordist bifurcation is embodied in the shift from an economy of quantity to an economy of variety, under the pressure of the relaxing of borders and increased competition (Veltz 1996). Hatchuel refers to this as a "Taylorism of variety" (Hatchuel *et al.* 2002, p. 32). Moreover, the flexible specialization model combines the advantages of the large enterprise (economies of scale) and those of the small business (productive flexibility), allowing for a better capacity for innovation (Garofoli 1992).

Literature on the district, and its variants, analyzes spatial concentrations of businesses that have formed on their own, based on family or neighborhood ties, or even with the aim of integrating the production chain of a territory. In this sense, Bagnasco evokes a "spontaneous social capital" specific to the Italian district[1], that is, "accumulated over the course of history within networks that define local society" (Novarina 2012, p. 21). He contrasts it with the "created social capital" that public authorities try to reproduce. Indeed, while the models we have just seen are the result of local initiatives, the spatial concentrations intended to rapidly bring science and industry together are the result of public policies, whether they be French technopoles or American science parks.

1 We recall that, in his study, Bagnasco emphasizes the structuring character of the Venetian parishes as places of information and exchange of know-how.

1.2. Spatial concentrations of technological activities

The second half of the 20th century brought about, on the academic side, mainly economic concepts of territorial innovation systems, as well as several policies promoting the bringing together of science and industry at a local level.

1.2.1. *The time of technopoles: reconciling regional planning and innovation*

In France, the technopole concept was part of the vast policy of decentralization that began in the 1950s and continued until the decentralization acts of the 1980s. The first attempts to create technopoles emerged in the 1970s, as in the case of the Meylan innovation and scientific and technical research zone (*Zone d'Innovation et de Recherche Scientifique et Technique*, ZIRST) in the Grenoble area and, especially, Sophia Antipolis near Nice, which remains the most striking French example. However, it was during the following decade that technology parks appeared all over France:

> Following the decentralization of 1982, the idea that development and coordination actions can promote the constitution of technological poles, and thus the economic development of cities or regions, spread among local elected officials, who were strongly encouraged in this by companies specialized in the development of business parks (Grossetti 1995, p. 3).

The state became involved in this development, initially through DATAR (*Délégation interministérielle à l'aménagement du territoire et à l'attractivité régionale*), and then gradually a multitude of operators emerged: mixed syndicates, associations, mixed economy companies, etc. As a result, in the 1990s, several works on technopoles were published, in an attempt to decipher the phenomenon and categorize the various experiments that were flourishing in France and abroad. For Jean-Yves Faberon, the emergence of technopoles in France is at the crossroads of three different policies (Faberon 1990): first, a regional planning policy (metropolises, new towns, conversion clusters and growth clusters resulting from the Guichard Report of 1986); second, an innovation and research policy (including the July 15, 1982, act aiming to develop regional technology clusters in which research and the socio-economic world are decompartmentalized for the development of technologies) and finally, a decentralization policy promoting the emergence of regional technology clusters. The state is thus strongly committed to the development of these work and production spaces, convinced by the concept of cross-fertilization (Savall and Zardet 2005). In *La fièvre des technopoles*, Jacques de Certaines (Certaines 1988, p. 28) identifies three types of technology clusters. First, desaturation clusters aiming to relieve metropolitan congestion: in France, Sophia Antipolis falls into this category; in the United States, Route 128 comes to mind in relation to MIT (Massachusetts Institute of Technology) and in Japan, Tsukuba in the suburbs of Tokyo. Second, conversion clusters operate in areas that have undergone major crises, for example, in the former mining basins of northern and eastern France, such as Meylan-Grenoble, Nancy-Brabois or Metz 2000. Finally, Jacques de Certaines observes a third type of technopole: development clusters as a deliberate strategy for developing a region through territorial anchoring of cutting-edge technologies (he cites the examples of Montpellier Languedoc-Roussillon Technopole and Rennes Atalante).

In their book, *Technopoles of the World: The Making of 21st Century Industrial Complexes*, Manuel Castells and Peter Hall analyze the different forms of industrial organization that technopoles can take (Castells and Hall 1994). Based on an empirical study of the principal global concentrations of technological activities, they distinguish between different types of technopoles, which show how broad the reference can be: high-tech industrial complexes (Silicon Valley, Route 128); science cities, where scientific research predominates, but where the link with the local industrial fabric is weak (the example of Tsukuba in Japan is highlighted); technology or science parks, which, for the authors, are the result of public initiatives for the economic development of the territory (Sophia Antipolis in France, Cambridge in the United Kingdom); even metropolitan areas such as Paris, Tokyo and London. From all these case studies, Castells and Hall identify three main motivations for technopoles: reindustrialization, regional development and the creation of synergies. The authors specify the long time needed to achieve these three objectives: 20–30 years to see the first effects.

Thus, for Faberon, the primary function of a technopole, and often the only one achieved, is to welcome business (Faberon 1990). In order to do this, various financial benefits are granted: free assistance services, notably through help in setting up projects, industrial property, interest-free loans, help in setting up a business through reduced rents and research and development subsidies. In addition, the site on which the technopole is located must be close to research and training institutions. It must also be well served by means of communication, proximity to highways, an airport, rail links, etc. Finally, as the quality of the buildings and the environment contribute to the image of the cluster, the site must have good landscaping and architecture and must include facilities (favorable living conditions, good health and school facilities, cultural life, etc.). The technopole is seen as a complete space combining three interrelated zones: an industrial fabric and work spaces, a core of institutes and research laboratories and residential areas for researchers and their families (Tatsuno 1987).

Thus, the technopolitan space is embodied in:

> A place of spectacular urban planning, high-end leisure facilities, and cultural events [...]. Indeed, creation is not limited to discoveries and applications in the scientific and industrial fields [...]. It is the sum of human activity which is carried along in the wake of the intelligence revolution. A revolution which, in order to bear all its fruits, must in fact encompass all spheres of existence, which supposes that intelligence does not remain limited to the economic field alone (Garnier 1988b, p. 169).

The technopole is therefore seen as a "new city in which the informal contacts generated by cultural, sporting and civic activities would have contributed to cross-fertilization" (Rasse and Arasziewiez 2007, p. 3). Nevertheless, some of the literature on technopoles, in addition to showing that the accumulation of all these factors does not necessarily guarantee innovation performance (Castells and Hall 1994; Cooke 2001), questions the artificiality of these spaces and the consequences for technopolitan workers. A detour through Japan and the work of Sheridan Tatsuno demonstrate the artificial dimension of the creation *ex nihilo* of the Tsukuba technopole (also studied by Castells and Hall):

> Many Tsukuba residents miss nothing more than walking through crowds on narrow sidewalks, sitting elbow to elbow in unimaginably tiny cafes and restaurants to warm their bodies and minds with sake and rich beers to maintain their long-standing relationships with the local butcher, baker, and

> police officer... It may be the city of brains, but it has neither heart nor soul (Tatsuno 1987, p. 139).

These critical comments on the artificiality of the technopole as a means of creating innovation spaces in areas that are, we might say, technology-free, occur early in the literature, but are still the minority view. Most of the literature on technopoles provides us with information on the mechanisms put in place, usually by public bodies, to generate a local culture, an innovation-producing environment. By highlighting a series of factors conducive to the creation of innovation, it differs from the literature on innovative environments, which seeks to determine how the "environment" is likely to innovate on the basis of the existing fabric (and in particular on the basis of a long-standing local culture, the history of the actors involved, etc.).

1.2.2. *A spontaneous and innovative environment conducive to a "technological atmosphere"?*

In the mid-1980s, regional and industrial economies differentiated between, on the one hand, "the regions that are winning", to use the title of Benko and Lipietz's book literally (Benko and Lipietz 1992), and on the other hand, those that are losing. The concentration, or otherwise, of research and development activities thus appears to be decisive. The European Research Group on Innovative Environments (*Groupe de Recherche Européen sur les Milieux Innovateurs*, GREMI) is investigating some 15 regions in Europe in a comparative approach in order to answer the question of why are some regions developing while territories with large manufacturers are in progressive decline. In this perspective, the analysis grid becomes the notion of environment. Linked to the life sciences, its definition is understood as "a set of external factors that act in a permanent or lasting way on living beings" (Coppin 2002, p. 32). Cangilhem emphasizes the predominance of the deterministic dimension in the notion of environment developed in biology (Cangilhem 2000). This systemic character is largely revisited in the concept of the innovative environment, insofar as the authors consider that innovation is no longer the product of single enterprises, but of the social, economic and political factors necessary for the innovation process. GREMI's authors combine an industrial analysis with a spatial analysis of innovation and base this on three axes (Camagni and Maillat 2006): the technological paradigm (the capacity of the enterprise to develop a technology with a view to differentiating itself from its environment), the organizational paradigm (the capacity of the enterprise to establish relationships with the other actors in its environment) and the territorial paradigm (the capacity of the enterprise to identify and take advantage of the attributes of its territory: know-how, competencies, as well as institutions, etc.).

One of the best known examples is the study of the Jura Arc (Maillat *et al.* 1992), which analyzes the transition from an industrial district centered on the watchmaking industry to a technological district focused on microtechnology. In this transformation, the role of the environment is highlighted insofar as innovation has taken place on the basis of the existing fabric: a shift from mechanical to electronic skills, the gradual integration of electronic products from watchmaking into printed circuit components, etc. (Pfister and Nemeti 1995, p. 29). In this regard, Benko tells us that innovative enterprises do not pre-exist in local environments, but are secreted by them (Benko 2007). Gradually, local training courses are adjusting the content of their teaching to this reconversion, such as the Lausanne Polytechnic, the Swiss Laboratory for Watchmaking Research and the Swiss Center for Electronics and Microtechnology. This adjustment is also favored by the environment, that is, the local industrial culture and the mobility of people on the labor market.

While Marshall spoke of an industrial atmosphere in his works on the district, theories on the environment tend to show that a technological atmosphere conducive to innovation can exist in some fashion within a territory. Among these works on the concentration of technological activities, divergences appear on the relevant spatial scale of such systems: from very localized business parks (Bernardy and Boisgontier 1988) to regional innovation systems (Cooke 2001) and urban innovation systems (Grossetti 2001).

Ultimately, all of these configurations can be understood according to the concept of "local innovation systems" (Gilly and Grossetti 1993), namely:

> A set of organizations (companies, research centers, universities, etc.) and individuals producing technological innovation on the basis of regular research and development activities within a given area (Grossetti 1995, p. 4).

The 1980s were therefore conducive to the research into the phenomenon of concentration of technological activities by social scientists, who linked it to the transition from industrial to cognitive capitalism.

1.2.3. *The era of cognitive capitalism: the race for creativity of individuals and territories*

Just as some observers attributed the emergence of industrial districts to a broader economic paradigm shift, some literature sees the changes of the 1970s and 1980s as a shift from industrial to cognitive capitalism, that is:

> In this new form of capitalism, the cognitive and intellectual dimension of work becomes dominant, and the central issues of capital development and some forms of property development are directly related to the transformation of knowledge into a fictitious commodity (Vercellone 2008, p. 72).

From then on, this economy would be based, not on knowledge, but on its exploitation, becoming, like information, a property in its own right. Moulier-Boutang, with reference to Karl Polanyi, believes that the new great transformation is embodied in cognitive capitalism, through global financialization and the digital revolution resulting from NICTs (Moulier Boutang 2007, p. 72). In this context, knowledge, creation and invention, which Moulier-Boutang synthesizes in the concept of immaterial work, are at the center of the mode of production. André Gorz, a philosopher in cognitive capitalism, draws up a non-exhaustive list of common goods that are ultimately usurped by the labor, privatization and commodification of this capitalism: biodiversity and common spaces, either urban or natural; genetic heritages; cognitive, cultural, affective and communicative resources of individuals; living knowledge in general, the product of cooperation and human interaction (Gollain 2010, p. 552). Thus, Gorz considers that cognitive capitalism is the crisis of capitalism. Having lost its relevant categories, capitalism seeks to surpass itself by exploiting an abundant resource, human intelligence, in order to produce scarcity (Gorz 2003, pp. 81–82). We have entered a capitalism that mobilizes competition through intensive innovation, in which the cognitive and creative capacities of individuals are at the service of an increased search for innovative products (Hatchuel *et al.* 2002, p. 32). It is therefore believed to have the power to counteract major trends, such as the relocation of industrial jobs or, to use the accepted expression, to slow down the "deindustrialization" of France (Guillaume 2008, p. 296) by converting industrial basins into valleys of knowledge. The logic of post-Fordist specialization and the production of scarcity linked to the development of cognitive capitalism are thus at the heart of local policies, which are seen as genuine "remedies for industrial decline".

In addition, in the early 1990s, there was a rapid spread of management theories from across the Atlantic, which developed a reading of the territory in terms of competition. They observed the effects of centripetal forces of agglomeration and at the same time centrifugal forces of international activities. Krugman was particularly interested in the differences between territories based on the positive externalities of agglomeration (Krugman 1991). At the same time, and in the same vein, Michael Porter developed the principle of comparative advantage, the keystone of his cluster concept. He transfers to territories an industrial logic of building assets in a competitive relationship within a market (Porter 1993). Specialization enables a comparative advantage to be put forward. In a capital mobility context, we are witnessing an explosion of competition between territories to attract the most qualified workers and businesses. Advantages are not given, but deliberately constructed (Carré and Levratto 2011, p. 360). From the point of view of economic development policies, this implies a shift from a territorial policy focused on local redistribution issues to a policy focused on externalities and attractiveness issues (Béhar *et al.* 2013, p. 20).

As early as 1989, David Harvey, in a critical view, drew attention to this shift from "managerialism to entrepreneurialism" in local policies. The "managerialism" of the 1960s was focused on services to residents, whereas "entrepreneurialism" is based on policies oriented towards an exogenous economic development strategy (Harvey 1989, p. 3). Harvey, too, places this orientation in the context of the recomposition of capitalism since the 1970s (the transition from a Fordist-Keynesian regime of capital accumulation to a regime of flexible accumulation), which has led to the development of interurban competition for resources, jobs and capital. This competition has been largely accompanied, even encouraged, by the state, notably through the logic of the call for projects, which acts as a means of territorial differentiation (Brenner 2004). By entering into this competition, local public actors have become veritable promotion agencies for their territory in order to attract investors, especially from overseas. The standard-bearers of this promotion are therefore the specific advantages that the territory possesses compared with its rival territories (Carré and Levratto 2011, p. 336). Beyond this mission of promotion, the public actor must take charge of the economic and social regulation that was previously assumed by the large Fordist enterprise that du Tertre and Laurent define under the term territorial regulation as:

> The capacity of a localized and institutional social system to guarantee accumulation processes in its territory, to maintain economic activities there, to develop and protect its natural resources and technological achievements and to guarantee social peace. Territorial regulation thus results from the necessity to regulate the conditions through which sectors are deployed and resources are allocated (Laurent and Du Tertre 2008, p. 15).

In this context, the public actor must encourage, mobilize, coordinate and structure interactions and resources (Béhar *et al.* 2013, p. 17). Thus, using the logic of Porter's comparative advantage, Anna Lee Saxenian questions the difference in development between the American West Coast innovation hub, Silicon Valley, and the East Coast hub, Route 128 (Saxenian 1994). By looking at the structure of the social networks of the two clusters, Saxenian opens the way to an abundance of literature on the role of geographical proximity, on the typology of commercial and scientific exchanges and on the place of non-market interindividual relations in the coordination of actors (Lanciano-Morandat *et al.* 2009, p. 178).

1.3. The valleys of knowledge: interindividual relations as a source of innovation

The end of the 20th century saw the emergence of a craze for valleys conducive to industrial and technological development, thanks in particular to the informal links forged by geographical proximity.

1.3.1. *Informal links in the heart of Silicon Valley*

Since Saxenian's work on Silicon Valley, the cluster concept is based on the idea that it can be crossed by interindividual relations that are conducive to informal exchanges, which are themselves conducive to the development of scientific and industrial collaborations (Saxenian 1996). The author explicitly refers to Porter's comparative advantage. Indeed, she attempts to uncover the advantage that Silicon Valley has over its Bostonian counterpart, Route 128. For her, the development of the Californian cluster is strongly linked to the structure of the social networks that cross it. Based on regional data, Anna Lee Saxenian notes that Route 128 created three times fewer jobs in the early 1990s than its Californian neighbor (Saxenian 2000). It was less efficient, particularly in terms of its volume of exports of electronic products and its ability to concentrate the most successful national companies. According to the author, economists neglected the role of social networks in the success of such industrial concentrations, and, for their part, sociologists implicitly accepted this neglect, reinforced by the idea that market processes are not suitable objects for sociological study, since social relations play only a minor role in them (Granovetter 1985). Thus, faced with two distinct industrial systems with similar characteristics (similar technology, set up after the war, etc.), Saxenian proposes an approach based on social networks in order to distinguish them and explain their different developments. For her, Silicon Valley is based on dense learning networks between actors from related technologies that encourage entrepreneurship and experimentation. Competition is intense, but it is supported by informal communication systems that create mutual adjustments and learning.

In the eastern United States, Route 128 is more composed of significant autarkical structures and the links between individuals are therefore characterized by a principle of corporate loyalty (Saxenian 2000). In the relationships between businesses, customers and suppliers, individuals refer first and foremost to the vertical hierarchy of the organization to which they belong, before circulating information horizontally between firms. For Saxenian, the performance of Silicon Valley, or its advantage, to use Porter's terminology, depends on local social and institutional determinants, such as professional or commercial organizations, far more than economic and fiscal factors (wage differentials, real estate costs, local taxes). For example, The Semiconductor Equipment and Materials International or The Software Entrepreneur's Forum organize meetings, dinners, seminars and training courses and participate in the creation of these social networks (Grossetti 2004, p. 170). From this point on, we can observe how interactions embedded in a given framework (in this case the businesses in the Valley) can be decoupled in another context (professional association, for example).

In another study of Silicon Valley, Michel Ferrary particularly observes the social networks of the French community living and working in the

Valley. These French expatriates meet in "non-economic social institutions" (the French school in Palo Alto, the Maison Française, the film club, French restaurants), ethnic celebrations (the July 14 national holiday, the Beaujolais Nouveau day), newspapers and associations with a more economic purpose (Doing Business in French, Eurotrash, Silicon French). They develop strong ties of friendship, support and appreciation through cultural and social activities during which they meet (Dibiaggio and Ferrary 2003). According to Ferrary, they mobilize this community in the economic activities in which they are engaged. We can therefore see that literature highlights the virtuous development of these territorial concentrations by emphasizing the effects of geographical proximity in the constitution of social networks.

1.3.2. *The relational logic essential to geographical proximity*

Since the beginning of the 1990s, the economics of proximity (Bellet *et al.* 1993, 1998; Rallet and Torre 1995; Gilly and Torre 2000; Dupuy and Burmeister 2003; Pecqueur and Zimmermann 2004; Torre and Filippi 2005) has sought to take account of the conditions necessary for the coordination of agents (Talbot and Kirat 2005, p. 9). This question has been the subject of research concerning the links, or otherwise, between geographical proximity and social interactions. The abundant work has led to "the emergence of a kind of specific scientific community, with its own conferences and regular publications" (Bouba-Olga and Grossetti 2010, p. 2). In the course of their work, this school has managed to define a certain number of types of proximity, but a concern for readability has led its authors to restrict the canonical forms of proximity to two (*ibid.*, p. 4). Thus, they agree to say that alongside geographical proximity is organized proximity. The former is doubly relative: on the one hand, it is relative to the means of transport and, on the other hand, it proceeds from the judgment of the individuals on the distance that separates them from other individuals or groups. The latter is not geographical, but relational in nature and is based on an organization's ability to have its members interact (Talbot and Kirat 2005). Beyond this vertical distinction, the authors have made a horizontal division within organized proximity between the logic of belonging (to the same structure, for example) and the logic of similarity (sharing common representations and a common system of values). The authors stress that the two proximities can be combined and thus facilitate knowledge sharing. Nevertheless, geographical proximity does not guarantee organized proximity, the absence of which translates into an extreme weakness of cooperative relations, solidarity links or shared representations (Torre and Zuindeau 2012, p. 354). Moreover, some authors postulate that spatial proximity may lead to negative externalities of agglomeration, such as industrial espionage, unbridled competition, local market saturation or the poaching of qualified personnel (Depret and Hamdouch 2009, p. 32).

Given the limits of geographical proximity (negative externality, absence of spontaneous links, etc.), other relational logics between scientific and industrial actors have been studied. Estades, Joly and Mangematin propose three relational logics (Estades *et al.* 1996): proximity, club and market. These three logics were later completed by Grossetti and Bès (2001), who then distinguished a "network logic" in which the contact results from a relational chain linking the two officials responsible for the collaboration. Most often, this relational chain has its roots in the university courses studied by the two officials. This network logic partially challenges the logic of proximity established by Estades, Joly and Mangematin, insofar as, for Grossetti and Bès, personal networks do not only stem from local proximity. They also identify "institutional logic", which is almost similar to the "club logic" of Estades, Joly and Mangematin, since both refer to the intervention of an external body, most of the time governmental, which provokes interactions between enterprises and laboratories. Finally, the two studies establish a third similar logic: that of the market. Estades, Joly and Mangematin consider that partnerships of this type are often the result of the will of industrialists who need solutions provided by researchers, whereas Grossetti and Bès consider that both scientific and economic actors can lead this market logic (Grossetti and Bès 2003, p. 343).

Ten years after the Grossetti and Bès study, Julien Barrier also identifies three major patterns of academic and industrial interaction (Barrier 2014). For him, the relationship is most often created on the initiative of industrialists who seek to: (1) enter into collaboration with researchers in order to acquire specific knowledge and know-how in a well-defined field, (2) optimize an existing technology that enables researchers to test theoretical models in an industrial context or (3) jointly explore, upstream of the development or application phase, concepts or research objects that are beneficial to the scientific production of academic researchers (Barrier 2014, p. 63). Geographical proximity is therefore gradually being supplemented by analyses that focus on other types of proximity and relational logic. Alongside these proximities, the sharing of a common social capital, which is decisive for establishing links, has also been the subject of several works.

1.3.3. *Social capital as a driver of innovation*

A number of works emphasize the role of individuals' social capital in facilitating the development of economic activities (Bourdieu 1980; Coleman 1994; Burt 1992). Coleman was particularly interested in the relationships and knowledge of individuals. In this approach, the social network becomes a resource in the same way as cultural or economic capital. In his works on Rational Choice Theory, he defends the idea that the individual determines his choices, not only on the basis of the economic calculation in terms of cost/benefit ratio but also on the basis of his relationships and knowledge. These authors do not question the idea that individual behavior is determined by economic incentives and financial resources. They add that they are also determined by motivations and social resources such as reputation, legitimacy and address books (Lanciano-Morandat *et al.* 2009, p. 180). This position is in line with the Bourdieusian thesis that capital must be accumulated in order to be fully operational. Another resource is then regularly evoked in the study of these social exchanges: trust. Social capital is not enough, and some authors argue that the condition for economic progress is the development of trust between people, understood to be the expectations nurtured by one individual towards the explicit or implicit promises of another (Dasgupta 2011, p. 50). Thus, in line with this work, some have analyzed the social capital of entrepreneurs as a decisive resource in the realization of a cluster (Feldman 2001; Feldman *et al.* 2005), considering that a network analysis of firms is, in fact, a network analysis of individuals:

> The firm is, in the best of cases, considered as a network of individuals, or as being akin to an individual. We remain in a universe where the active force is the individual and where the collective is constructed or reduced to a set of relations between these individuals (Callon and Ferrary 2006, p. 39).

Further to the main distinction made by the proximity school between a geographical and organized dimension, Michel Grossetti and Olivier Bouba-Olga propose a new decomposition of socio-economic proximity by an approach at the level of individuals. They distinguish between two types of situations: proximity of resources and proximity of coordination, for which, first of all, resources are cognitive ("in the heads of the actors"), and they evoke a plethora of notions used in the social sciences to describe the behavior of actors from the inside: "strategies", "dispositions", "affects", "habitus", etc. The authors specify that these resources are shareable between them (language, values, standards) and can be mobilized to coordinate (Bouba-Olga and Grossetti 2010, p. 8). The resources are also material and social proximity is played out at the level of the resources possessed (assets, income, qualifications, social status, etc.). Alongside the proximity of resources, they mention the proximity of coordination, which is made up of both mediation mechanisms and social networks. They cite the often used example of the resources mobilized in the labor market: individuals rely as much on intermediaries (job applications, classified ads, recruitment agencies, etc.) as on personal relationship networks (word of mouth). Grossetti believes that these chains of interindividual relations are the most empirically founded of the various theories describing proximity effects (Grossetti 2004, p. 168). Indeed, many case studies highlight the existence of local interindividual social networks that cross organizational boundaries.

Another view also relies on the social capital of individuals but underlines the relations between individuals and the advantages created by the constitution of a network. A significant proportion of literature is based on the concept of embedding in the study of concentrations of industrial and/or scientific activities. This notion was developed by Granovetter in a celebrated article on the relations between principals and subcontractors in construction companies (Granovetter 1989). Within these interactions, he highlights the fact that the actors are personally involved in the economic relations they maintain. This personal dependence between individuals forms the notion of embeddedness. Relationships between enterprises cannot be understood without paying attention to the personal ties that unite the individuals involved in these interactions. Granovetter adheres to the tradition of social network analysis, which considers that interactions are the only acceptable starting point for sociological analysis. However, according to Grossetti, this approach tends to reduce economic activity to these sets of social relations (relational reductionism), when he considers that:

> The social structure is not reducible to a network, it also includes groups (or circles, fields, worlds, etc.), in other words collective entities endowed with boundaries of name, affiliation procedures, etc (Grossetti 2006b, p. 85).

Consequently, alongside the notion of embedding, Grossetti and Bès use the notion of decoupling developed by Harrison White (1993). While he is one of the founders of the network approach, White also considers that it can easily lead to a naturalization of relationships, leaving aside other types of social structures: families, organizations, groups, etc. Therefore, embedding always refers to the dependence of an actor on the personal ties he has with others, and decoupling is, conversely, the self-nomination of this actor with respect to these relationships in a new embedding process. In this context, Grossetti and Bès speak rather of network dynamics in order to evoke the two complementary processes of embedding and decoupling. On the question of where relationships originate, Grossetti and Bès rely on the work of Claude Fischer (2011), who, based on an empirical study during the 1970s, tells us that most individuals meet through structures deemed stable and lasting (family, work, neighborhood, school) and that brief encounters in a bar or restaurant rarely convert into strong ties. Thus, when we look at the genesis of relationships, the notion of embedding works in reverse, with individual relationships embedded within collective structures. Nevertheless, they may become autonomous from those frameworks from which they originate, which is called decoupling. Grossetti and Bès indicate that this process "can be perfectly transposed to firms whose exchanges are initiated in a given framework (a "market") and can unfold in other frameworks" (Grossetti and Bès 2003, p. 53). There is an abundance of this literature on social networks in economics, sociology, management sciences, etc., which particularly feeds the concept of the cluster as a product and producer of informal exchanges conducive to innovation.

2

The Management Roots of the Cluster and Its Worldwide Dissemination

The cluster concept, as it has been applied at the local level in France, is the product of mostly management literature, the authors of which are located between academia and supranational decision-making bodies, particularly the OECD and the European Union. Through the paradigm of the knowledge economy, the promotion of American success stories that convey concepts such as "coopetition" or categories such as "knowledge workers", these researchers provide policy makers with a theoretical and practical toolkit for the development of clusters. The concept has been transposed into public policy. Starting in the United States and promoted by international organizations such as the OECD and the European Union, the cluster is gradually being imposed into the national framework that has been promoting links between science and industry for the past 30 years.

2.1. An economic and management concept destined to become a public action mechanism

While the innovation cluster was being conceptualized by economics and management researchers, its dissemination within public policy in Western countries was achieved in a short time, virtually simultaneously.

2.1.1. *Porter's cluster: the rapid spread of success stories*

The cluster concept became popular in 1990 with Harvard University (Boston) Business Strategy professor Michael Porter. In *The Competitive*

Advantage of Nations, he defines a cluster as "a geographical concentration of companies and associated institutions, interconnected within a particular field and linked by common elements and complementarities" (Porter 1990). At first glance, this definition suggests that the cluster is a rediscovery of established evidence of the district, technopole or innovative environment concepts (Torre 2006, p. 17), but differs in that it is organized and instituted. Strengthening the links between cluster actors is a policy objective of Porter's work, which advocates public intervention in the constitution of these local clusters (Renaud 2015, p. 67).

Moreover, unlike the work on districts, clusters do not refer only to groups of enterprises in the same sector but also to research structures, as was also the case in the work on innovation environments and technopoles. Nevertheless, in his initial approach, Porter describes intense internal competition as well as essentially vertical coordination processes (with subcontractors, suppliers, etc.) along a value chain. Productivity, through cost reduction, appears to be the priority issue, ahead of innovation, which assumes a more horizontal coordination, enabling the transfer of knowledge, primarily between laboratories and enterprises. In 1998 and 2000, Porter clarified his definition by emphasizing the combination of cooperative and competitive relationships between firms, underpinned by formal or less formal social networks, all of which lead to better learning and diffusion of innovation (Leducq and Lusso 2011, p. 7).

Porter's approach has been widely accepted, and two dimensions have been particularly examined: territorial anchoring and market-oriented relationships. Indeed, there are works that raise the question of the relevant geographical scale of the cluster (Cooke 2001) and focus on the relationships maintained within the cluster and not outside it. At the same time, following on from Porter, clusters are analyzed from the perspective of transactional and contractual relationships (Depret and Hamdouch 2009, p. 23). Based on the work of Nalebuf and Brandenburger, management professors from Yale School of Management and Harvard Business School (Nalebuf and Brandenburger 1996), Gordon and McCann apply the principle of coopetition to cluster relationships. By contracting the terms "cooperation" and "competition", this oxymoron, which is no longer an oxymoron in the authors' view, refers to the long-recognized phenomena of alternation between states of competition and cooperation (Blanchot and Fort 2007,

p. 166)[1]. Gordon and McCann therefore consider the relations between actors in a cluster from the perspective of coopetition (Gordon and McCann 2000). In management and economic literature, there are also a series of authors who clarify Porter's concept and emphasize the innovation (Leducq and Lusso 2011) or high-tech cluster (Depret and Hamdouch 2009). In particular, Achermann sees "a triple helix that turns at full speed to accelerate the emergence of innovations" (Achermann 2014, p. 12). These works identify the regional scale as relevant, as it would allow for alternation between global and local resources (Veltz 1996; Cooke 2001), and assess that innovations require a sufficient "critical mass" of skills and therefore a greater need for coordination than in less R&D-intensive sectors.

Based on a statistical study of the evolution of the number of scientific publications on the subject of clusters, Suire and Vicente demonstrate the rapid diffusion of the concept. By researching contributions in economics and innovation policy analysis, they observe a clear upward trend between 1998 and 2008, followed by a period of stability up to the present day. Beyond the metric dimension, the authors inform us that, in terms of content, the concept of networks as "market or non-market relationships of economic actors who participate in a collective process of innovation and knowledge exchange" (Suire and Vicente 2015, p. 95) runs through this literature. The main common thread in this work is the idea that, without the networking of actors on a local scale, the potential for innovation remains limited. Progressively, clusters are perceived according to an approach that is no longer exclusively competitive, but more reticular (Leducq and Lusso 2011, p. 22).

The cluster concept is all the more popular as it appears at the heart of so-called "success stories". Indeed, some studies consider that the idyllic vision of the cluster, which is so popular with many academics and policymakers, needs to be qualified, since it is analyzed from the perspective of successful examples[2]: "Silicon Valley, Route 128, San Francisco Bay Area, Research Triangle Park, Italian districts, Baden-Württemberg, etc."

1 In the 2000s, in France, a typical case of coopetition was the agreement on the price framework in the telecommunications sector between Orange France, Bouygues Telecom and SFR.

2 "Success" factors for a cluster are generally considered to be: the industrial or commercial success of start-ups that have become large companies in a market, the development of a technology with global repercussions or the financial benefits for the site.

(Depret and Hamdouch 2009, pp. 8–9). This focus on success stories and the strong tendency of literature on links between universities and industry to focus only on success stories (Renaud 2015) are accentuated by the emphasis on the cases of Stanford University and MIT (Owen-Smith 2003). These two examples hold a significant place in the literature describing the economic and territorial development of Silicon Valley and Route 128 in the United States (Saxenian 1994) and contribute to a positive and pacified image of cooperative relations between universities and businesses:

> While there are many works on these two successes, there is little or no trace of the failures, i.e., the more numerous cases where the relationships between universities and business, on which regions or municipalities had gambled, did not produce such fruitful results (Trépanier and Ippersiel 2003, p. 77).

Indeed, based on a corpus of works on the relations between research institutions and business between 1995 and 1999 in scientific journals and those published by governmental or international organizations (OECD), Trépanier and Ippersiel show that the objective is often the same. They are almost exclusively success stories, more than half of which are located in the United States (37 out of 67), followed by Europe (but only 22 mentions, including 8 in England) and the remaining 9 in the rest of the world. The sample is over-represented by English-speaking countries, which implies industrial and university systems with state-of-the-art equipment and scientific and financial resources; indeed, large companies and high-tech SMEs occupy a predominant place. On the academic side, these are institutions with national and even international reputations, with significant scientific output, rather than regional universities focused on training and with less celebrated research teams (*ibid.*, p. 78). The corpus highlights a series of success stories whose initial ingredients remain exceptional: prestigious research and scientific production, large industrial groups, state-of-the-art equipment and a market facilitated by an economic liberalism specific to English-speaking countries. In this literature, the model most often put forward is that of Silicon Valley. Its success is said to have been made possible by a relational density built up over the years (Suire and Vicente 2015):

> From the 1960s to the present day, California, more than any other place on the planet, has brought together the scientific and cultural factors favorable to the networking of individuals and

> to the proliferation of techno-social innovations that have resulted (Dagnaud 2016, p. 15).

This narrative is also based on the idea that Silicon Valley and the advent of the internet are the result of a counterculture comprising hackers and hippie communities whose project was, above all, utopian rather than economic (Flichy 2001). The enthusiasm for these "cultural creatives" (Ray and Anderson 2001) of Silicon Valley should be qualified by the historical weight of public investment. The Internet, or even GPS, on which GAFAMs[3] proliferate today, relies not only on these "garage innovators" but also on large U.S. federal state funds, from government organizations and the military (Sainsaulieu and Saint-Martin 2017, p. 13). Moreover, the literature places little emphasis on the economic failures experienced by the cluster of excellence, particularly during the dot-com bust, which wiped out about 127,000 jobs (9% of regional employment) in the space of 18 months and half of the gains accumulated over the period between 1998 and 2000 (Leducq and Lusso 2011, p. 18). Silicon Valley thus found itself in a situation of over-specialization or mono-activity (Torre 2006, p. 30). By focusing on the nascent Web industry, it was at the mercy of either a change in the competitive universe or a financial collapse. Furthermore, it is interesting to mention that one of the founding fathers of Silicon Valley, former Stanford University vice president Frederik Terman, was asked by Bell Labs to replicate the California Valley experience in the supportive environment of New Jersey, but never succeeded (Duranton *et al.* 2008: 19). Although the process of imitation and replication seems risky, it is clear that much of the literature on clusters has rarely been analyzed beyond the success stories (Torre 2006, p. 18).

2.1.2. *Knowledge management and its workers as a dominant paradigm*

Porter's cluster is also part of a series of models designed for public action. The most famous are the triple helix (Leydesdorff and Etzkowitz 2000) and the *New Production of Knowledge* (Gibbons *et al.* 1994). The triple helix refers to an imaginary figure of dynamic interdependence between the three spheres of industry, the state and science, and conveys the

3 GAFAM is an acronym for the giant companies of the Web: Google, Apple, Facebook, Amazon and Microsoft. They are also known as the Big Five.

idea of a social order increasingly based on knowledge (Shinn 2002b, p. 25). Knowledge is thus generated through fluid relationships between the state, markets and science. The model does not differentiate between the academic, industrial and governmental spheres, but this configuration of three interdependent and equivalent helices is the product of a slow and progressive historical process. For the authors of the *New Production of Knowledge*, it is more of a rupture. They show how we have moved from Mode 1, characterized by a closed academic world with watertight disciplinary fields, to Mode 2, structured by interdisciplinarity and interactions with industry. The thesis of Gibbons, Novotny, Limoges, Schwartzman and Trow has been successful in the English-speaking and Nordic countries by proposing the simple opposition of the old university, where researchers asked and answered questions, to the new, where society poses problems and researchers, as experts, provide solutions (Schultheis *et al.* 2008). In his conclusion on the collective work *Le cauchemar de Humboldt. Les réformes de l'enseignement supérieur européen (The Humboldt nightmare. European higher education reforms)*, Yves Winkin emphasizes the prescriptive character of Mode 2, compared with the more descriptive Mode 1: "Today, there is no longer a colloquium of the OECD, the World Bank or the European Union on 'knowledge management' that does not use the contrast between Mode 1 and Mode 2" (Winkin 2008, p. 200).

At the dawn of the 2000s, the drive towards the neoliberal "knowledge-based economy" was thus at the heart of international bodies such as the OECD and the European Union (Bouchez 2016, p. 33). One of the first references to this now-used term is an OECD report, *Employment and Growth in the Knowledge-based Economy*, coordinated by economists Foray and Lundvall (Foray and Lundvall 1996). They highlight the correlation between the so-called knowledge-based sectors of activity and economic growth. Their thesis is the result of a series of seminars organized by the OECD. They developed the idea of a knowledge economy as a set of activities oriented towards the production, distribution, exchange and consumption of goods and services linked to intangible goods and/or cognitive capacities linked to knowledge. The authors have subsequently supplemented the OECD report with several publications (Foray and Lundvall 1997; Foray and Cowan 1998; Foray 2000; Lundvall and Archibuji 2001; David and Foray 2002). In these works, it is mentioned that current economies are directed towards knowledge production and no longer towards industrial production, which has been left to the economies of emerging countries. This postulate therefore implies the emergence of a new economic discipline. With reference to the development of the industrial economy during the 1820s, at the same time as the advent of large-scale industry, so the knowledge economy is emerging at a time when economies are progressively based on knowledge production:

> The knowledge-based economy corresponds essentially, in each country, to the production and service sectors based on knowledge-intensive activities. These are usually identified by combining indicators of knowledge production and management, such as research and development (R&D) expenditure, the employment rate of graduate workers and the intensity of use of new information technologies (Foray 2009, p. 3).

In order to define these knowledge-intensive activities, the authors rely on the earlier work of the economist Fritz Machlup, who includes education, communication activities, data processing equipment, information services and other information-related activities in this sector. In 1962, Machlup showed that this industry accounted for 29% of GDP in the United States in 1958 (Machlup 1962). Foray takes up this typology and completes it by affirming that knowledge-intensive activities proliferate in other production and service sectors.

The paradigm of the knowledge economy is more a particularly operative discursive category than an analytical framework. Indeed, it is becoming a watchword that makes innovation a requirement, "a major factor of economic competitiveness that tends to become the almost unique way to survive and prosper in highly competitive and globalized economies" (David and Foray 2002, p. 14). Additionally, within the framework of the knowledge economy, much of the economic and management literature at the end of the 20th century considers innovation to be the result of information and knowledge sharing within networks (Badillo 2013, p. 25). Indeed, although the sectoral basis of the knowledge economy is indeed based on science and technology, it also develops thanks to territories that have been able to set in motion a dynamic of agglomeration of the resources characteristic of this economy, namely highly qualified personnel, research and development laboratories and innovation services (Foray 2009, p. 5).

The knowledge economy is thus based on another concept derived from management sciences: the category of "knowledge workers". Works on the post-industrial era (Touraine 1969; Cohen 2006; Rifkin 2012) show the tertiarization of the economy and the increase of service workers in finance, health, administration, etc., and refer to these types of workers. However, it was one of the fathers of modern management, Peter Drucker, who first used the expression *knowledge workers* in 1957. He defines this category as individuals who apply what they have learned in the development of concepts, ideas and theories, rather than through manual skills. Also, according to Drucker, unlike manual workers, organizations should not tell these knowledge workers what to do, but provide them with the right conditions to emulate, apply their knowledge and ultimately increase their productivity (Drucker 2010). Robert Reich defines the knowledge worker as a symbol analyst, in the sense that these workers create or manipulate symbolic representations, particularly with computers (Reich 1997). In a similar vein, Jeremy Rifkin, a specialist in economic forecasting, speaks of "abstraction handlers" capable of using computers and networks to identify and solve problems (Rifkin 1997). In this management literature, the emphasis is on the economic role of these workers: they are the ones who create value based on their knowledge and creative work. Knowledge workers' main activity is to sell their intellectual output to customers (Bouchez 2004). However, this definition excludes a great many individuals who also produce knowledge, such as teachers or librarians, for instance.

Whether we adopt a narrow definition (limited to the commercial or technological spheres) or a broad one (from doctors to artists via financiers), these knowledge workers are often characterized by a particular way of working and their lifestyles. These dimensions can be found in the rather vague concept of the creative class developed by Richard Florida. Alongside the working class (which is said to be shrinking and which only accounts for 20% of employees) and the service class (in which he groups fast-food workers, cleaning and security companies, office workers, etc.), which is growing fast (45% of the working population), there is also a growing group of individuals who use creativity in their work, which he calls the creative class (one-third of the working population) (Florida 2002, p. 235). According to the author, the challenge for public decision-makers is to attract this class by investing in cultural facilities of interest to them (theaters, museums, trendy bars, cultural events, sports facilities).

In France, the work of Pierre-Michel Menger takes up the concepts developed by Florida. He too considers engineers, artists and scientists as "the hard core of a creative class or an advanced social group", whose creator appears as "the exemplary symbol of the new worker" (Menger 2003, pp. 7–8). Menger particularly dwells on the artist who, through the precariousness and instability of his employment, represents the symbol of the professional of the future, that is, inventive, mobile, unresponsive to hierarchies, motivated, caught up in an economy of the uncertain and more exposed to the risks of interindividual competition and the insecurities of professional trajectories. A number of criticisms have been leveled at this and particularly Florida's work: the correlation between cultural life and economic dynamism, and in particular the example of Montreal, which is very dynamic culturally, but whose economic growth remains weak, is often put forward in order to testify to the weakness of the argument (Vivant 2006); the simplistic notion of social class, which in reality is only an amalgam of intellectual occupations (Durand 2016); or the statistical construction and indicators used to evaluate the presence or absence of the creative class (Levine 2004). The most recurrent criticism remains the broad nature of the general theory. This criticism is not unique to Florida, who, in spite of this, has and continues to be invited to lecture to a number of local public policy makers. Overall, much of this literature, especially management, seems to use theoretical nesting that remains vague.

2.1.3. *A theoretical and practical toolkit provided by researcher-experts in clustering*

The literature on clusters has, and continues to enjoy, some success in the management and economic academic community, as well as among public policy makers, as Torre describes very well in the following excerpt:

> The centrality of geographic clusters has often remained far removed from the understanding of mainstream economists [...]. We had to wait for the work [...] on economic development processes, which focused on the phenomena of spatial concentration of innovation or research activities. This is Porter's work (see, for example, Porter 2000) on clusters, which has had an even greater impact than Krugman's, since it is not limited to the economic discipline alone and directly influences the principles adopted by policy makers at both local and national levels. Clusters are now considered as the basis for local and even national policies in many countries (UK, Germany, Netherlands, etc.). For example, they have been used

> as a basis for the recent reflection on 'Local Production Systems' in France and can be compared to the new 'Competitiveness Clusters'. More surprisingly, they are often considered as major development tools by the major operators of the globalized economy – see OECD 2001 and 2005, or World Bank 2002 (Torre 2006, p. 16).

Indeed, whether referring to local and national policies or international organizations, the cluster seems to give "turnkey" solutions to political leaders. This is the observation made by Shinn in his analysis of the work on the triple helix, in which he shows that public decision-makers must work on finding small variations in the concept (Shinn 2002, p. 28). Business leaders, politicians and administrators are invited to constantly rethink their directions, through, as Shinn tells us, symposia (including three international symposia on the triple helix) (Amsterdam 1996; New York 1998; Rio 2000) and publications related to these issues, in order to help decision-makers in the conduct and application of the concepts. Regarding the cluster, it would seem that the recipe is as simple as relying "on the resources of its regional territory, located at the intersection of the local environment and global forces" (Leducq and Lusso 2011, p. 23). Indeed, the authors add that "all territories, including the least developed, have their chances, whether they are old industrial regions or emerging countries" (*ibid.*). This capacity for declination relies on the deliberately vague character of the cluster, which enables it to easily adapt to different local situations. Martin and Sunley rightly show that the vague and incomplete definition of the Porterian cluster has greatly contributed to its popularity, as it can be used in a wide range of cases and interpretations. Nevertheless, they consider that it is precisely the elasticity of the notion that makes it a "chaotic concept", since its universalist character can be a potential source of ambiguity and amalgam:

> However, although the definitional and conceptual elasticity of the cluster concept can be seen as a positive strength, in that it enables a wide range of cases and interpretations to be included, we consider it to be problematic. The concept has acquired such a variety of uses, connotations and meanings that it has, in many respects, become a 'chaotic concept', in the sense of conflating and equating quite different types, processes and spatial scales of economic localisation under a single, all-embracing universalistic notion (Martin and Sunley 2003, p. 71).

As a result, we can see that, worldwide, political institutions, the European Union and national and local governments advised by their experts (OECD, World Bank, regional development agencies, etc.) are implementing clustering policies. As part of this process, Porter, Gibbons, Florida, Novotny, etc. are consulted, even hired, by these different political bodies in order to apply this global trend: "Clusters, it seems, have become a worldwide fad, a sort of academic and policy fashion item" (Martin and Sunley 2003, p. 4). These decision support experts are frequently referred to as "management gurus", who come to the rescue of private managers and public institutions (Bruno and Didier 2013, p. 72). They are located in, or on the frontier of, the academic world and enjoy a significant audience within policy arenas, offering a "scholarly package with a practical purpose" (Sainsaulieu and Saint-Martin 2017, p. 14), for which innovation is the raw material. This metaphor leads Saint-Martin and Sainsaulieu to identify them as "packaging manufacturers"[4]. This figure of the researcher–expert is similar to that of the secant marginal, in that it is:

> A stakeholder in several interrelated systems of action who can, as a result, play an indispensable role as an intermediary and interpreter between different, even contradictory, logics of action (Crozier and Friedberg 1977, p. 86).

Richard Florida is the typical example of the "celebrity consulting professor" (Sainsaulieu and Saint-Martin 2017, p. 15) who has managed to convert his research into a lucrative private speaking activity (about $50,000 per conference), as reported in *Le Monde* on April 10, 2009, which also uses the term "guru" in its headline: "Richard Florida, controversial guru of urbanism (*Richard Florida, le gourou controversé de l'urbanisme*)"[5]. He has been invited to many Western cities to deploy his methods of attracting the creative class, which have enjoyed a significant rate of propagation within

---

4 In the introduction to *L'innovation en eaux troubles. Sciences, techniques, ideologies*, Ivan Sainsaulieu and Arnaud Saint-Martin emphasize French personalities such as Jean Tirole, the 2014 Sveriges Riksbank Nobel Prize winner, and Philippe Aghion, Chair of Institutional, Innovation and Growth Economics, at the Collège de France. They are both presented in the media as "prophets of innovation", helping public authorities in the conversion of "the entrepreneurial state at the heart of the globalized paradigm of the innovative society" (Sainsaulieu and Saint-Martin 2017, p. 14).

5 Allix, G. (2009). Richard Florida, le gourou controversé de l'urbanisme. *Le Monde*, April 10 [Online]. Available at: http://www.lemonde.fr/planete/article/2009/04/10/richard-florida-le-gourou-controverse-de-l-urbanisme_1179144_3244.html.

international, European and then national political discourse. At the same time, the work of Drucker and Porter advocated that public policymakers move away from fixing market failures to fixing network failures; the "arsenal of public incentives" thus developed a series of mechanisms in this sense, for the majority of cluster variants in most developed countries, with the aim of bringing together individuals with high innovation potential (Suire and Vicente 2015, p. 95). The vagueness of the concept makes the cluster a notion with different uses, interpretations and examples of application (Largier *et al.* 2008, p. 22). Its lack of robustness has lead, as we have seen, to academic criticism, "while the political world is quicker to seize on it because it 'speaks' better than the usual economic theories" (Lamy and Le Roux 2017, p. 94). Indeed, the concept has been widely taken up at global, European and national political levels. We will also see that the figure of the researcher–expert is at the heart of each policy level.

2.2. Global dissemination of good clustering practices

The knowledge economy, from which the cluster stems, has a considerable influence in industrialized countries. We can thus observe a worldwide homogenization movement based on this paradigm.

2.2.1. *A paradigm born in the United States and forged at the heart of the OECD*

French research and innovation policies are often justified by overseas examples, most often from North America (Louvel and Hubert 2016). The success stories mentioned above are indeed becoming exemplary and should be reproduced in other contexts. Louvel and Hubert have clearly shown the influence of major international paradigms on nanoscience and nanotechnology science policies. A similar influence is at work with the knowledge economy paradigm, insofar as public R&D policies in industrialized countries have reinforced the process of contextualization of science, essentially by bringing it closer to industry (Brunet 2011). The United States has been a leader in the construction and diffusion of this paradigm. As early as the 1970s, biotechnology appeared to be a source of economic development. In 1975, the Asilomar conference was held in California, bringing together researchers in molecular biology and life sciences. It aimed to anticipate, in the context of biotechnology development, the industrial, economic and social consequences of precise intervention on the genome of living organisms.

After 20 years, the accession of the Clinton and Gore Democrats to the summit of American government led to the implementation of a complex economic, financial and legal mechanism for transferring scientific resources from the public to the private sector (Uzunidis 2007, p. 54). The traditional science and technology policy (major research and development programs in the energy, defense, space sectors, etc.) was replaced by a research and innovation policy. The latter is oriented as much towards public research programs as towards business (Branscomb and Keller 1998). In this context, American economists close to the Democratic Party conceptualize the processes of transfer from one business to another, from public institutions to private entities. Avoiding mechanisms of domination and power relations, they see it as "multiform, multifunctional and multipartner cooperation" (Uzunidis 2007, p. 65).

The OECD, along with the World Bank, is one of the international institutions most involved in the dissemination of the concepts of the knowledge economy paradigm, notably by supporting the policies of the United States. In a 2006 report, the institution highlights the actions of industrialized countries that: (1) favor funding for research carried out under the supervision and control of industry, (2) reform their university systems to make them more competitive in order to improve the supply of scientific and technical services to companies and (3) promote the mobility of researchers and their involvement in business (OECD 2006). At the same time, the institution points out that certain obstacles to complete efficiency remain, such as the civil servant status that researchers often enjoy, or the evaluation of public research, which is still based on the criterion of published work and not on the contribution of researchers to industry. As early as the 1980s, the OECD, based on the US case and with the help of economists, developed the normative and neoliberal concept of national innovation systems (Lamy and Le Roux 2017, p. 70).

One of these liberal economists, John Stanley Metcalfe, defines the concept as a set of institutions that contribute to the development and diffusion of new technologies. These institutions form the framework within which government implements policies to foster the innovation process. It is a system of interconnected institutions that create, store and transfer knowledge and skills and define new technologies (Metcalfe 1995). The cultural authority of economics and the prestige of the OECD give legitimacy to all of these precepts as set down in national policies (Lamy and Le Roux 2017, p. 71). Moreover, policies to link science and industry are not only at work in the highly industrialized OECD countries. Indeed, developing countries do not hesitate to reproduce policies in vogue in more developed countries and call on experts in the field. In Asia, in particular, Michael Porter and Christian Ketels have produced a series of works on the implementation of a Cluster Mapping Project in Thailand, whose objective is to strengthen the competitiveness of the region. Using Porter's methods, this organization is assessing the possibilities for creating and supporting cluster formation in Thailand (Leducq and Lusso 2011). This broad global dissemination is made possible because:

> It comes from an organization that has managed to establish itself as an essential "interpreter", but also because, as an indicative, non-binding guide to good practice used by default (there are not so many in "model" markets), it leaves room for the actors who refer to it to legitimize their own political action in practice (Lamy and Le Roux 2017, p. 72).

In a context of neoliberalization of public administration, the OECD manages, through strategic documents and management tools, although not legally binding, to influence national policies.

2.2.2. *To adopt OECD recommendations, or have them imposed?*

In their book, *Benchmarking, l'État sous pression statistique* (Benchmarking, the state under statistical pressure), Isabelle Bruno and Emmanuel Didier explain that, in the context of new public management, the OECD is developing a genuine benchmarking policy (Bruno and Didier 2013). Beforehand, in their introduction, they tell us that this managerial method means "evaluating by comparison with a model, a standard, an external norm" (*ibid.*, p. 9). In a globalized market, enterprises must no longer simply produce more, but must offer the best. In this context, benchmarking is the cornerstone of the panoply of instruments devised by management experts. This method was popularized by Robert Camp, an executive at Xerox, in 1989 to describe the identification of good practices among competitors in order to improve the company's performance. Developed in private companies, benchmarking has been transposed to public organizations and the liberal slogan of "less government" has been replaced by the neoliberal injunction of "better government" (*ibid.*, p. 10). Using this logic, the OECD is an organization created to produce expertise and recommendations for member states, thus aiming to standardize public policies in terms of both content and form. As early as 1994, the institution produced a document entitled "Performance Management in Government: Performance Measurement and Results-Based Management" in which government officials were urged to move "from a culture of rule enforcement to a culture of performance" (OECD Public Management Committee 1994). After two years, in 1996, in its policy paper The Knowledge-based economy, the OECD stated in its foreword:

> Identifying 'best practices' for the knowledge-based economy is a focal point of OECD work in the field of science, technology and industry. This report discusses trends in the knowledge-based economy, the role of the science system and the development of knowledge-based indicators and statistics (OECD 1996, p. 3).

This excerpt is consistent with the idea that the OECD, lacking supranational competence and even less democratic legitimacy, would have focused on its sole source of power: the collection of "data on its richest and most industrialized member countries with a view to producing comparable statistics to support its recommendations" (Bruno and Didier 2013, p. 114). In this way, it becomes able to influence states by comparing and ranking them (knowledge-based indicators and statistics) and ultimately by forcing them to adopt the best practices it identifies and recommends. In 2014, the institution went so far as to recommend that tuition fees be increased in fields that would offer (again based on statistics) the best job opportunities (Granger 2015, p. 73). In the case of clusters, the OECD disseminates its recommendations through cluster policy guidelines (Suire and Vicente 2015, p. 97), in which we find, in addition to statistical overviews, a good proportion of the management literature mentioned above, which praises the Porterian logic of clusters and the imperative of densifying networks as a condition for their performance. In these guides, public intervention is considered to "boost the production of innovation through synergy effects that could not exist in a context of organizational isolation" (Suire and Vicente 2015, p. 97).

More generally, we observe a diffusion of these management concepts through the role of "carriers" (Winkin 2008) that certain academics have played between the scientific, economic and political worlds, either mandated or directly within global political institutions. The French academic world has paid little attention to these very Anglophone debates, and the dissemination of their concepts has mostly been done away from the academic world (Winkin 2008; Lamy and Saint-Martin 2017). In contrast, they have been at the heart of political arenas. We recall that Foray and Lundvall forged their concept of a knowledge-based economy in OECD seminars. As for the authors of Mode 2, their thesis has progressively been imposed as a flagship work, like a charter for public, scientific and private decision-makers concerned with ensuring value creation through innovation policies based on science and its high degree of dependence on other spheres, particularly industrial (Brunet 2011). This appropriation and success in Europe is all the more rapid because, at the time of the publication of Gibbons, Nowotny and Scott's book, *Re-Thinking Science: Knowledge and the Public in an Age of Uncertainty* (Gibbons *et al.* 2003), Helga Nowotny was a member of European bodies: from 2001 to 2005, she chaired the European Research Advisory Board (EURAB), which played an advisory role to the European Commission. It is in this context, and by figures at the frontiers of science, economics and politics, that the European higher education system will be imagined through the Lisbon Strategy and the Bologna Process.

2.2.3. *The European Union, sponsor of the race to the knowledge economy*

Convinced by the strategies of its transatlantic neighbor, and above all in order to compete with it, the European Union launched into the race for the knowledge economy at the beginning of the 21st century, to make its territory the global nest of knowledge and ideas. The Lisbon Strategy, decided by the European Council in March 2000, aims to make the European Union "the most competitive and dynamic knowledge-based economy in the world". From that point on, higher education appears to be the fundamental instrument at the service of the economy and leads to quantified and rational logics based on the education rate:

> Estimates suggest that a one-year increase in the average educational attainment of the population translates into

increased productivity growth rates of 5% in the short-term and an additional 2.5% in the long-term (Council of the EU 2004).

The Lisbon Strategy is a tool for intervention by the European Union to unify and standardize national university systems. Isabelle Bruno's work highlights the major significance of European decisions in the higher education and research area (EHERA) (Bruno 2008). The author analyzes the Lisbon Strategy as the imposition of a neoliberal frame of reference on national systems. She shows the influence of a diffuse and non-regulatory process of Europeanization, through practices such as benchmarking (see section 2.2.2.), which invites university actors to be more competitive, in opposition with legally binding European directives. Thus, a series of major normative guidelines is being put in place and disseminated in the universities of the Member States, which Christine Musselin has grouped together under the term "scripts" (Musselin 2008).

The first of these five imperatives is "to modify the role of the state", principally by disengaging it from university affairs through the creation of new bodies for the management and evaluation of higher education policies (AERES, ANR, etc.) leading to a policy of resource allocation and accreditation according to performance (*ibid.*, p. 16). The second script is "to transform universities into organizations". In France, this process of autonomy was achieved through legislation relating to university responsibilities, known as the LRU (*Loi relative aux Responsabilités des Universités*). The most striking example of this conversion of universities into organizations is the management of positions and people, which is no longer the responsibility of the state. Universities therefore develop rational hiring relationships and control procedures, similar to those between employers and employees. The third script is "to increase the role of stakeholders", with the idea often taken up by public decision-makers that the university must "come out of its ivory tower". The minutes of the European Council of March 20–21, 2003, call for "the stimulation of interactions between industry and research institutions if we are to realize our potential to create new businesses" (Granger 2015, p. 69). Here, we find the "New Knowledge Production" thesis, on the need to make research socially relevant, moving to Mode 2 of knowledge production. The fourth script is "to enter into the logic of privatization". Universities are encouraged to "redouble their efforts to transform ideas into real added value" by adopting a funding system similar to that of the United States (European Council of March 20–21, 2003, *ibid.*). These funds "may derive from sponsorship, research contracts with companies or public services, or from student fees" (Musselin 2008, p. 19). On the one hand, these funds would come from:

> Enterprises that could, for example, be invited to finance or co-finance equipment, schools, scholarships, curriculum renewal activities, university chairs or departments, research universities, training courses to attract students and apprentices to fields facing a shortage of skilled workers (Council of the EU 2003, p. 17).

Derived from the private sector, these techniques bring employer investment into the university, which in turn guides teaching and research and creates a competitive situation between universities to find the best investors and financial resources. On the other hand, one of the sources of funding for higher education should come from users. This second resource requires an increase in tuition fees presented as a rational "cost/benefit" calculation for the student. It is shown as vital that the latter must invest in studies in order to reap a benefit in future years on the labor market (Garcia 2009, p. 161). The Lisbon Strategy, with its dissemination of these principles of university financing, intends:

> To produce economic growth through investment in human capital and technological innovation, but also to instill independent thinking into students regarding economic utility, of which the European institutions are the guardians (Garcia 2009, p. 163).

Finally, the last script is "to place oneself in a global perspective". This quickly brings to mind the Bologna Process and its objective of harmonizing European systems in order to facilitate and intensify exchanges. In 1998, the French Minister of Education at the time, Claude Allègre, brought his German, Italian and English counterparts together to sign the Sorbonne Declaration. This declaration calls for the emergence, alongside the euro and economic treaties, of a Europe of Knowledge that includes the recognition of qualifications obtained in the European university system. On June 19, 1999, this process took shape with the famous Bologna Declaration, which was signed by a larger number of countries (29 European states, both EU and non-EU members) than a year earlier in Paris. European leaders committed to the creation of a European Higher Education Area (EHEA) based on the two fundamental principles of the Sorbonne Declaration, namely the creation of a university system organized in three cycles (bachelor's degree, master's degree, doctorate) and the implementation of the ECTS (European Credit Transfer System) as standards, in order to harmonize the various European degrees. This process was thus implemented during the 2000s, punctuated every two years by conferences of the signatory countries to monitor the progress made.

The Lisbon Strategy and Bologna Process thus invite universities, newly responsible for their own funding, to move closer to industrial actors and to adopt a mode of organization that originates from the private sector. Moreover, the management logic is that the regions should be "mobilized by Europe and the state in order to promote entry into the knowledge economy" (Crespy 2007, p. 27). We then observe competition between territories under the guise of territorialization of research and innovation policies, in which clusters will be mobilized (Laperche and Uzunidis 2010). Again using the logic of benchmarking, the European Union is setting up "the European Cluster Observatory, whose objective is to permanently monitor the dynamics of the most innovative clusters in order to establish a common methodology for the European Union" (Leducq and Lusso 2011, p. 16).

If we look at the cases of the OECD and the European Union, we can see that international organizations are both places where the legitimacy of ideas and their relays for dissemination are constructed (Godin 2002). As a result, science and technology policies in France are not immune to this movement of European appropriation of the Anglo-Saxon model of the relationship between science and industry. This interweaving is now considered irreversible and has become essential to change national modes of public action in order to adjust to the globalized environment (Branciard 2002, p. 34).

2.3. The French legislative framework from the 1980s to the 2010s: a favored coming together of science and industry

In this globalized context, from the 1980s, France adapted its legislative framework to the paradigm of the knowledge economy. The main changes affected the academic world, which was considered by legislators to be too far removed from economic issues and not well adapted to the labor market.

2.3.1. *Researchers converted into entrepreneurs*

Beginning in the 1980s and especially during the 1990s, France also worked to disseminate strategic science (Rip 2002), that is, "basic research conducted with the expectation that it will produce a broad base of knowledge that can be used as a basis for solving recognized practical problems, whether current or future" (Irvine and Martin 1984, cited in Peerbaye 2004). France made its first legislative breakthrough with the Act of July 15, 1982 on "the orientation and programming of research and technological development in France" (Vergès 2010, p. 17). For private law professor Étienne Vergès, by placing the discovery of new results and their dissemination on the same level, the Act introduces a paradigm shift that reduces the scope of the ethical principle for researchers. With this legislation, research organizations are authorized to set up a research commercialization service, to create development companies or even to create subsidiaries. From a statutory point of view, researchers cannot carry out public research and private commercialization activities at the same time. However, they can be seconded to a company to carry out research, development, technology transfer or knowledge transfer activities. They can also be made available for the creation of an innovation company and to then return to the public service once their task is completed (*ibid.*, 2010).

We then observe a penetration of private legal tools within public research that shakes the scientific ethos based on the Merton model: universalism, communalism, disinterestedness and organized skepticism. The Act of December 23, 1985 on research and technological development did not create new legal tools, but extended the 1982 Act by placing the commercialization of research at the heart of the budgetary objectives of the new national research policy. Ten years later, the Interministerial Committee for Scientific and Technical Research (*Comité interministériel de la recherche scientifique et technique*, CIRST), meeting under the chairmanship of Prime Minister Alain Juppé on October 3, 1996, declared:

> While underlining the need to maintain a sustained effort in favor of fundamental research, the CIRST considers that the research system should be more oriented toward meeting socio-economic objectives. It also emphasizes the need to commercialize the results of public research.

The text then insists on the necessary transfer of technology towards "socio-economic objectives". The 1999 legislation on innovation and research, known as the Allègre Act, enshrined this principle. It is directly inspired by the American model, in the sense that it intends to use scientific innovations as factors for economic growth. Claude Allègre, then Minister of Research, presented the legislation as a necessity:

> Let's take a look at U.S. growth in recent years. About a third is driven by the new information and communication technologies and more than half by the high-tech industries (statement by Claude Allègre, June 3, 1999).

This speech is part of a rhetoric developed a few years earlier by Claude Allègre, notably in a book published in 1993, *L'âge des savoirs. Pour une renaissance de l'Université* (The knowledge age. To revive universities). In this book, Allègre delivers his vision of the French university, behind the United States, as mentioned in the above quote, and marked by its "viscosity" and inertia (Allègre 1993). He explains that the French university needs a renaissance, which must involve adapting its teaching to professional needs and opening up to the global circuit of capital (Granger 2015, p. 63). In 1999, he therefore presented the Innovation and Research Act as a way of dusting off the old university machinery and the resistance that was forming within it. This legislation sets three essential objectives: researchers with public status are encouraged to create a company or to participate in its development; the creation of innovation companies is advocated; the transfer of public research results to companies is similarly supported. Thus, thanks to measures that derogate from the principles of the civil service, which require that civil servants devote their entire professional activity to the tasks entrusted to them, research personnel can start a business as a manager or partner. It also authorizes collaboration contracts or research consortiums between the public and private sectors. It was in this context that the first public incubators were created to provide researchers and other project leaders with the necessary financial and methodological support to set up businesses and commercialize their research, using a variety of legal forms, primarily public: internal university services, private companies, associations, public interest groups, scientific interest groups, etc. In 1999, most operators had already created their own development structures, which explains this diversity of legal forms (Vergès 2010). The turn of the century saw the emergence of the first French companies created by researchers with public status and the consolidation of their support structures.

In addition to legislative and legal tools, France has set up seminars similar to those of the OECD and the European Union. Thus, in the first quarter of 2001, the General Planning Commission organized a "seminar of experts on the knowledge economy [...] constituting the preliminary stage of a reflection [...] on the theme of France's integration into the knowledge economy" (Paillard 2001, p. 6). Among these experts were academics[6], representatives of the Ministries of Economy, Finance and Industry, Employment and Solidarity, National Education, the European Commission and research institutes (in international relations, on qualifications, etc.). The seminars addressed the various themes of the knowledge economy, as conceived in supranational bodies: the spatial polarization of innovation activities, questions of cooperation and coordination in the knowledge economy, the measurement of links between private and public actors, the mobility of young doctors, to name but a few (*ibid.*, p. 5). The experts formulated their conclusions based on a theoretical corpus similar to the management literature previously mentioned (Foray, Lundvall, Florida, Machlup, etc.), which helped to disseminate them within French institutions. In the same vein, in 2004, the Blanc Report was submitted to the Prime Minister. This document, in the wake of the Allègre Act and its public incubators, served as a guide for the implementation of competitiveness clusters policy, which grants subsidies to groupings of laboratories, training programs and businesses in a given sector, on a regional scale.

2.3.2. *The university: a link in the cluster supply chain*

In his report to the Prime Minister, *Pour un écosystème de la croissance* (For a growth ecosystem), Deputy Christian Blanc prioritized the competitiveness of territories, with the aim of remedying the social and economic handicaps that prevent France from achieving sufficient growth. He suggested developing synergies between laboratories, enterprises and training at the local level, in a competitive context. The same year, a DATAR report entitled *La France, puissance industrielle. Une nouvelle politique industrielle par les territoires. Réseaux d'entreprises, vallées technologiques, pôles de compétitivité* (France, an industrial power. A new

6 The academics Lundvall and Foray participated in these seminars; however, Foray was no longer presented as an economist but as the representative of the OECD. It would seem that he had resolved his dual membership in the academic and political world by wearing his expert hat and abandoning that of a researcher.

industrial policy through territories. Business networks, technology valleys, competitiveness clusters) made the same observation as Christian Blanc, with both referring to the work of Michael Porter. As an example, Blanc made sure to recap on the cluster management concept:

> The cluster model, which Michael Porter defines as "a group of companies and institutions that share a common field of expertise, are geographically close, connected, and complementary" is actually at the heart of economic development [...]. The people who work in these institutions meet, discuss, understand and are able to engage with each other, thus enabling the development of clusters (Blanc 2004, p. 13).

The Blanc Report emphasized the regional level as relevant for the emergence of clusters and proposed several solutions for transferring a certain number of responsibilities for regional planning and the management of incentive aid to companies to the regions themselves (Jacq 2011, p. 7). This logic of regionalization of the economy also aimed to give more weight to universities by creating regional clusters, at the expense of national research organizations. Nevertheless, the state remained the main organizer of this policy, which, following a national call for projects, selected and accredited 71 competitiveness clusters. A budget of 1.5 billion euros was allocated over three years (grants for collaborative projects and exemptions for researchers' positions in companies). Clusters were, for the most part, characterized by large, leading corporations, central to the cluster's interaction system (e.g. Airbus for Aerospace Valley in the Midi-Pyrénées region and Sanofi for Medicen in the Paris region). While the state's historic concern was to avoid a concentrated and unequal economic geography by supporting declining territories, it is now encouraging competition between territories in order to obtain accreditation and encouraging specialization and cooperation within the most dynamic territories in order to create interactions that are favorable to innovation (Duranton *et al.* 2008, p. 12). The 2009 report by senators Michel Houel and Marc Daunis gives a "unanimously positive assessment" of French clusters (Houel and Daunis 2009, p. 1) and adds that "the competitiveness cluster policy is in line with the Lisbon Strategy" (*ibid.*, p. 65). Indeed, the reports make sure to emphasize that this is a global dynamic that should be both followed and encouraged. The Blanc and DATAR reports provide many positive examples of clusters in Europe (Catalonia, Denmark, etc.) and in the rest of the world (Australia, Brazil, etc.).

The competitiveness clusters policy was subsequently reinforced by the National Strategy for Research and Innovation (*Stratégie Nationale de Recherche et d'Innovation*, SNRI) from 2007 to 2012, which principally lead to the creation of thematic research alliances to coordinate research and development priorities in defined sectors. In the case of biotechnology, this includes the National Life Sciences and Health Alliance (*Alliance nationale pour les sciences de la Vie et de la Santé*, AVIESAN). The LRU Act has greatly increased the autonomy of universities (as well as the harmonization of courses based on the European LMD model), and their competition has been facilitated by the creation of financial (National Research Agency, *Agence Nationale de la Recherche*) and evaluation (Agency for the Evaluation of Research and Higher Education, *Agence pour l'évaluation de la recherche et de l'enseignement supérieur*) agencies. Based on the prescriptions of the Bologna Process, as is the case in other European countries (Vallier 2011), France is adopting a research system with reduced training cycles, whose content must first of all enable the employability of students and correspond to the criteria of an evaluation agency (AERES) in order to obtain, at the local level, the sufficient credits granted by a national agency (ANR). Under the guise of autonomy, universities are not, ironically, free to define their training programs, constrained as they are by the liberal logic of evaluation and calls for projects:

> By assigning an imperative of competitiveness, competitiveness and performance to institutions, it [the LRU Act] has established the conditions for their managerialization based, on the one hand, on internal budgetary rationalization and steering by objectives and, on the other hand, on successive calls for projects that have enrolled them in a relentless race to obtain the means for their survival. The Eldorado promised by the transition to "expanded responsibilities and competencies" (*Responsabilités et Compétences élargies*, RCE), which was supposed to free universities from the shackles of the state, has, in fact, placed eight of them in a state of "supervised autonomy" under the supervision of the rectors, while the others are now experiencing the joys of seeking their own funding. Canvassing companies, begging for donations from alumni networks, increasing tuition fees, in short, "selling out", these are the new skills that universities have gained (Bruno and Didier 2013, p. 194).

In this context, university presidents are trying to meet government requirements, while at the same time pulling their weight in an increasingly competitive world, by partnering with the local industrial fabric (Granger 2015, p. 87). The frame of reference is, in effect, that of the knowledge economy in which the university becomes a link in the "supply chain" within clusters (Lamy and Le Roux 2017, p. 103). Following the various reforms initiated by the European Union at the end of the 2000s, a policy of bringing research centers and private firms closer together was established in higher education by the EHERA (Faure 2005). The clustering policy was not limited to competitiveness clusters and took several forms. Indeed, it implied, as advocated by the Blanc Report, a multilevel dependence by the academic community on local private actors (Lamy and Le Roux 2017, p. 103). Indeed, the cluster concept was then taken up by local public authorities.

2.3.3. *A cluster for every territory*

Seduced by the political scope of competitiveness clusters, local authorities support, or are at the origin of, structures dedicated to innovation. This phenomenon has been reinforced since the 2010s. There are no longer any cities, regions or even departments that do not promote a local cluster. Presented as an antidote to relocation and job shortages, we can cite, in no particular order: Amiens Métropole, whose economic policy is based on three clusters (e-health, digital and energy autonomy); the city of Lyon, which is launching its "bio-district, a melting pot of biotechnological innovation"; or the Centre-Val de Loire region, which, through a call for projects, has selected a dozen clusters in several fields. The competitiveness cluster model is being used by local authorities "on the lookout for new ways to ensure growth and strengthen the competitiveness of their territories, regardless of their initial level of development" (Leducq and Lusso 2011, p. 15). Incubators that predate the Allègre Act are gradually being renamed as clusters. This is the case of Genopole, which, on the eve of the 2010s, gradually set aside its biopark name to identify itself as a biocluster. The institutional landscape of clustering attempts is therefore very varied. Groups of companies, public incubators and technology clusters are almost all listed by the *Association France Clusters*, whose board includes directors of several clusters (including competitiveness clusters). As a service tool for clusters, the association offers its members legal monitoring, partnership research, methodological support, strategy guides, conferences and even training. We can thus observe a form of institutionalization of territorial innovation systems.

Recently, alongside clusters, we have seen a dynamic of territorialization around spaces of a different kind: third places. It is difficult to define this fashionable concept, which can be called anything from coworking spaces, to project accelerators, to digital fabrication laboratories (the famous "fablabs"). For the most part, these are devices for coordinating and activating individuals with high technological potential (Suire and Vicente 2015, p. 111). Indeed, what distinguishes them from clusters is their ability to bring together individuals (not organizations) in an enclosed space, usually a large hub or several offices (not in the same territory) and to incite temporary proximities (Torre 2008). These places are usually frequented by mobile, highly educated, freelance or even remote workers. It is not surprising that a plethora of these places have initially sprung up in large capitals (New York, London, Berlin, Barcelona, etc.). In the space of a decade, the Paris region has seen the emergence of spaces of this type with evocative names, always recalling a relaxed and convivial spirit: the Hive, the Canteen or the Anticafé, to name but a few.

Although most of them are developed in the field of computer science and new information and communication technologies, biotechnology is also an integral part of this continuous cognitive research/technology/application of science in a logic of territorialization (Branciard 2002). Indeed, the first fablab specialized in biology was also created in the Paris region, at La Paillasse. Initially installed in a garage in the inner suburbs, it joined the majority of other third places in the Paris' Second Arrondissement in 2014. The phenomenon is spreading to other large French cities, supported by local authorities. Symptomatic of a new way and vision of working, free from the supposedly historic shackles of enterprise (fixed working hours and places, hierarchy, procedures), these places are the factory of a new type of worker, who is free to create and undertake. Boasting a culture of "tweaking", and a subversive spirit inherited from the hacker movement, they actually embody the heroes of unbridled capitalism (Lallement 2015, p. 400). They criticize the cluster concept and the concrete forms it can take (Martin and Sunley 2003; Torre 2006; Duranton *et al.* 2008; Hamdouch and Depret 2009), or the theories associated with it (Levine 2004, Vivant 2006 and Durand 2016, on the creative class). Nevertheless, despite these criticisms and proven cases of failure, the concepts of the cluster and the creative worker continue to be relayed by now-celebrated carriers from the academic and/or expert world to the political spheres.

Thus, international recommendations have direct implications on the French legislative framework and consequently on local policies. An isomorphism (DiMaggio and Powell 1983) can be observed, in which each state and/or each territory adopts clustering policies. Moreover, the logics of mimicry between territories lead most clusters to set up similar coordination and networking mechanisms (Suire and Vicente 2015, p. 114). The next chapter focuses precisely on the appropriation of the cluster concept in a particular territory.

The establishment of Genopole, created in 1998, is directly linked to that of the French Muscular Dystrophy Association (*Association Française contre les Myopathies*, AFM) in Evry. In the 1980s, the AFM gradually succeeded in initiating a scientific policy in the Evry area, which intensified with the creation of its own laboratory, Généthon. This acquired international renown with the mapping of the human genome and put the AFM at the forefront of the scientific scene. In addition to academic recognition, it can count on popular support with the organization of the Telethon, which punctuates the French associative landscape at the end of each year. The association thus relies on a financial manna, which few French laboratories can boast, and on a worldwide scientific legitimacy, which gives it an unquestionable territorial power. Its president then imagined an extension of the territory with the "Généthon Industries" project, surrounded by state-of-the-art infrastructures. It was therefore necessary to convince the public authorities to install the National Sequencing and National Genotyping Centers, which were necessary for the arrival of industrialists onto the site.

It was in this context that in 1998 Claude Allègre, then Minister of National Education, Research and Technology, entrusted Pierre Tambourin with the mission of creating the first biopark in Evry, dedicated to genomic and post-genomic research and to the development of the health biotechnology industry. The mission letter entrusted him with the support of local players the AFM and the University of Evry, which became his founding members and which would subsequently need to focus more on biology education. The creation of the biocluster was therefore the result of a project by the French government, which wanted to not fall further behind in the biotechnology sector, in view of its international competitiveness, and to "draw its credibility from a strong belief in the benefits of the American cluster model" (Brunet and Dubois 2012, p. 399).

Initially set up as an association under the 1901 legislation, the biocluster has been managed by a public interest group (PIG) since 2002, along with a semi-public company that manages the site's development. In addition, the two central departments of the PIG, Genopole Entreprises (GE) and Genopole Recherche (GR), are strengthening their respective missions of business creation and laboratory development. Through a policy of accreditation and service provision, the biocluster has gradually succeeded in bringing together science and industry players into its area: 86 biotech companies (around 60 of which are located on the Evry campus), 19 research laboratories, 24 shared technology platforms[7] and life science teaching structures. Initially focused on the installation of laboratories, we have observed over the last 10 years that the number of employees in companies is greater than those of laboratories, in connection with an intensification of training in entrepreneurship for researchers. In fact, in 2005, the staff in accredited laboratories represented 49% of the cluster's workforce, compared wih 33% who worked in business. In 2016, the trend was reversed: 51% of staff came from the business world versus 37% from academic laboratories (the remaining staff were divided between infrastructures and administrations).

With the development of its accredited companies and laboratories, the biocluster intends to impose its own territoriality on already established structures (AFM, the National Sequencing and National Genotyping Centers, the University of Evry), or on structures being set up outside the accreditation process, such as the Southern Paris Region Hospital (*Centre Hospitalier Sud Francilien*, CHSF) in 2012. Genopole has to deal with multiple territorialities and affiliations. As a result, clustering leads to the confrontation of divergent logics that are juxtaposed and coexist, but without necessarily creating interactions. In this context, the biocluster then sets up a series of communication and territorial marketing instruments alongside an intermediation device that are the subject of the second part.

Box 2.1. *A local example of a specialized life sciences cluster model: Genopole*

7 The figures are from the end of 2016.

3

The Cluster Imaginary: Tools, Local Narrative and Promise

In this world of success stories, innovation awards and dissemination of good practices, cluster administrators at the local level show that they, too, are part of the game. Thus, based on a particular cluster, this chapter reveals the imaginary that is locally conveyed to public and private financiers, as well as to laboratories and businesses, to encourage them to set up there. To do this, the cluster studied uses performative instruments such as benchmarking, territorial marketing and visual instrumentation techniques; a narrative based on a particular local history that is conducive to scientific research and industrial development; finally, a series of promises linked to employment in the territory and to the innovations enabled by the cluster. All of these resources act on the representations of individuals and lead to the construction of a sociotechnical imaginary (Jasanoff and Sang-Hyun 2009), which, through the rhetoric of innovation, images and tables, promises a desirable future.

3.1. Performative instruments: benchmarking, territorial marketing, visual instrumentation

The biocluster's Communication Department is setting in place a range of tools that showcase a high-performance, dense and collaborative biocluster through rankings, dashboards, diagrams and visuals.

3.1.1. *Benchmarking or territorial mimicry*

We have just seen how supranational institutions, such as the OECD or the European Union, are implementing a comparison and ranking system in order to highlight the good practices identified in their neighbors (or competitors) and to use them as a guide for the way forward. During the survey within the biocluster, these comparative analysis practices were observed on several occasions. Indeed, benchmarks are regularly carried out on other innovation sites in France in order to find exemplary clusters in the Kuhn sense (*exemplar*), that is, "solutions to concrete problems, accepted by the group as paradigmatic, in the very usual sense of the term" (Kuhn 1990, p. 397). It is a matter of finding an organization or territory with better performance or greater success, listing the means of achieving the same result, submitting them to decision-makers and evaluating them (Helgason 1997). The benchmarking technique, identifying and proposing the application of the most effective "recipes", has become a form of convergence thinking (Carré and Levratto 2011, p. 371) and a powerful relay of imitation by recommendation. This convergence logic functions from a decontextualization that makes the seemingly ideal model applicable everywhere.

At first glance, this technique does not seem too restrictive, since it calls on incompletely defined and non-coercive formal frameworks ("soft regulation") (Lamy and Le Roux 2017, p. 106). The strength of benchmarking does not lie in the diktats of management, but in highlighting the indisputedly best results it has recorded elsewhere (Bruno and Didier 2013, p. 104). It thus directs behaviors towards a mode of management from the private sector, to the rescue of "dusty" bureaucratic hierarchies, exacerbating competitive spirit and rivalries with other clusters. This import is all the simpler to carry out as we are witnessing a profound transformation of personnel in the public sphere. Indeed, there are more bridges than before between the civil service and the private sector.

The newcomers to public administration have had international careers, often moving into auditing and consulting firms, or into industry, moving more than their elders between the private and public sectors. These more hybrid careers have enabled them to familiarize themselves with new management techniques (*ibid.*, p. 11). They are also recruited for this knowledge, as is the case, for example, for the coordinator who initiated the benchmarking in question. She has a doctorate in biology and initially had experience in research laboratories before training in marketing and working for a large pharmaceutical group in the market research and competition departments. She was able to put her experience in the private sector and her knowledge of marketing techniques to good use at Genopole. As we shall see, most of the employees who work for pharmaceutical companies have a dual background in the public and private sectors.

Although this regulation is soft and the benchmarking guidelines appear not to be particularly coercive, this does not mean that they are not applied effectively. In January 2017, the report submitted to the Regional Council that presented the new organization of the biocluster moreover emphasized the "context of strengthening regional competition in terms of incubators/nurseries". In this context, from 2017 onwards, Genopole has been converging towards forms of organization inspired by other incubators, such as NUMA, in central Paris. In the Evry model, the incubator is called Shaker (Rise for NUMA) and the accelerator becomes Booster. The functions of the two facilities in Paris and Evry are exactly the same: they are aimed at young entrepreneurs (Genopole places particular emphasis on doctoral and post-doctoral students, inspired by NUMA's School Lab), accompanied by mentors, coaches or buddies, depending on the name, from neighboring companies.

This standardization of innovation spaces is reminiscent of the standardization that is taking place in global cities. Take the example of port cities that promote the artistic development of their docks: Barcelona today resembles Baltimore or Belém (Fournet-Guérin 2008). In France, Marseille is also trying to imitate its Spanish neighbor, which it takes as a reference in its urban policies.

3.1.2. *Territorial marketing: asserting the cluster's symbolic capital*

Paradoxically, while a standardization is largely at work in clusters or among innovation actors, at the same time each tries to highlight its own assets, most often through indicators and statistics, in particular through the production of dashboards (Bruno and Didier 2013, p. 28). These tables have a dual function: they serve as a support for decisions and the evaluation of objectives and provide a showcase to the outside world for the health of the organization. In a competitive and fund-raising context, the biocluster therefore produces annual activity reports, whose relevance depend on the indicators selected. Indeed, these cannot be exhaustive and take into account all available figures. Certain indicators are therefore selected for the figures to be both monitored and drawn attention to. In the report published in 2017, reviewing the 2016 results, there is a graph in the section entitled "Genopole contributes to the rise of French biotechnology" that presents the statistics in a positive light. At first glance, we learn that in 2016, the cluster handled 114 projects, had 644 contacts since its creation and 275 files appraised. Of the 114 projects, there were in fact 15 companies that would have joined the site in 2016, 5 files under study, 23 prospects (potential project leaders with whom Genopole were in discussion or negotiation) and, the majority, 71 companies accredited before 2016. In this respect, the biocluster also highlights the fact that 180 companies were accredited between 1998 and 2016, but does not report the total number of companies that have left the biocluster (56), mergers and takeovers (9) and those that have ceased trading (29), which represents a total of 94 companies out of the 180 accredited.

By means of these indicators, clustering involves building attractiveness on the model of a commercial brand (Lamy and Le Roux 2017, p. 107). In the same vein as benchmarking, clusters use territorial marketing policies that have been proven in the communication implemented by cities. In France, the city of Lyon was one of the pioneers, launching its anagram-shaped brand "ONLY LYON". In the same way, Genopole has become a brand (Genopole©) that awards its accreditation to the organizations that join it. These accreditation practices are imported from the logic of territorial specialization, which gives a particular value to a place, as is the case with AOCs (*Appellation d'Origine Contrôlée*) (Harvey 2008). In a similar way, accreditation guarantees a quality specific to its location, where the advantages are linked to the concentration of infrastructures and the symbolic capital attached to these high places of global capitalism (Clerval 2011, p. 180). Indeed, clusters are very often understood to be places that are highly intensive in knowledge and excellence: "a cluster of excellence in life sciences", according to the 2016 activity report (p. 31). The aim is to make companies want to be identified as innovative in the context of global competition. This is evidenced in particular by the Association of University

Research Parks (AURP), which in the fall of 2018 awarded Genopole its "Research Park of Excellence" accreditation. This example is then revealing of the top-down logic of the global paradigm of the knowledge economy towards clusters. Accredited by the American proponents of the concept, the biocluster in turn accredits companies and laboratories.

Here, accreditation has the particularity of having both a material and a symbolic function. We will see in the next chapter that it enables its members to benefit from the cluster's services (support, infrastructure, etc.), provided they are geographically grouped with the other accredited members. It therefore goes hand in hand with a power of territorialization. At the same time, it symbolically identifies different organizations as belonging to a common designation, a guarantee of innovation and performance. Indeed, the accreditations, whether they contain the terms "silicon" or "valley" (Suire and Vicente 2007), are based on symbolic capital or geographic charisma (Appold 2005). This symbolic capital is not only generated by the accreditation, but also by a series of images that highlight a dense and coherent campus.

3.1.3. *Visual instrumentation: the image of a dense, expanding campus*

In a study of progressive institutes, Pollock and Williams show how classification work, as well as the use of images, enables organizations to shape their discourse (Pollock and Williams 2010). Sainsaulieu and Saint-Martin, building on the work of Pollock and Williams, describe the toolbox of these organizations to strengthen their arguments:

> Quantitative instruments and the use of numerical data, visual instrumentation such as schemas, plans, designs and photographs, or even cognitive instrumentation, such as the use of shared values or reinforcing the legitimacy or credibility of the spokesman (Sainsaulieu and Saint-Martin 2017, p. 13).

As we have just seen, Genopole uses classic techniques from *new public management*, such as benchmarking and territorial marketing. Through these quantitative instruments and numerical data, both private and public institutions use visual instruments to create shared representations. Indeed, the works on the imaginary concept all emphasize the mental significance of symbolic images and their impact on individual behaviors and social frameworks. Castoriadis gives as examples architecture, urbanism, festivals, or even the means of communication organized according to an imaginary which does not cease to inhabit the individuals or the groups and which in return feeds this imaginary (Castoriadis 1999). For Castoriadis, without imaginary, there is no society. It is the utopias and representations which provide a bedrock of common beliefs and which structure the social link. Imaginary is an essential component which cements society, because all of these representations form a common world of meanings for the individuals who act within it. For Gilbert Durand, the myth is considered as a cultural formation interpreting social life (Durand and Cazenave 1996) and, as for Edgard Morin, *Homo sapiens* is also *Homo demens*. Starting from an anthropological analysis of the cinema, for the latter, imaginary life enriches and organizes reality (Morin 1958). These works all emphasize the link between the strength of mental and physical images and social reality. Thus, if we take a closer look at the images disseminated by the cluster, we realize that they give a very particular representation of reality. For example, in order to give the impression of a vast integrated campus where laboratories and companies coexist, communication documents disseminate images that highlight a dense and homogeneous geography, in perpetual expansion.

This echoes economist Paula Stephan's survey of the U.S. research system, which shows that space dedicated to biomedical and health research structures increased by 30% between 2001 and 2011 and that square footage is becoming a performance indicator (Stephan 2015) (see Figure 3.1).

We can see the progression of the biological, biomedical and health sciences curve, which has grown from 40 million square feet in 1988 to 90 million in 2011. Occupying space appears, visually and physically, to be a guarantee of success and growth.

Genopole was the first biocluster to be created, and this precursory aspect is at the heart of the narrative put forward by the cluster. The narrative is based on three structuring dimensions that draw on both historical and current elements: Evry as a place intrinsically linked to genomics research, the economic development of the region and the need to be globally competitive.

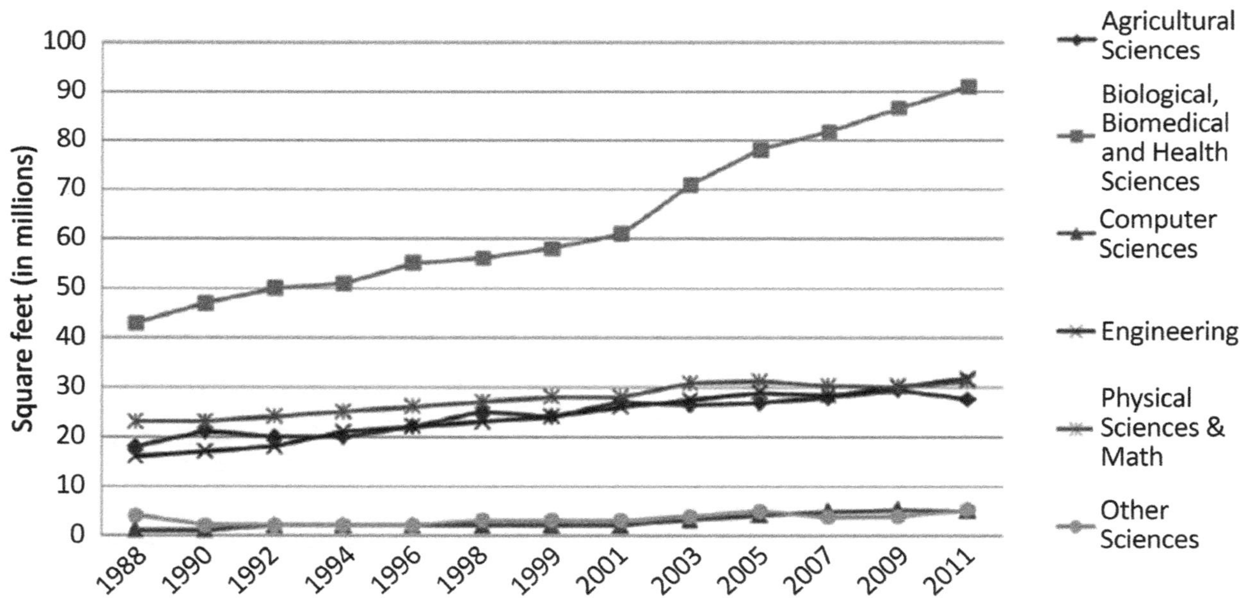

Figure 3.1. *Changes in square footage dedicated to research structures in the United States (source: National Science Foundation, 2013, https://voxeu.org/article/us-university-science-shopping-mall-model)*

3.2. The construction of a narrative

The performative instruments are thus supported by a rhetoric conveyed inside and outside the cluster, by the site's leaders, as well as by local and national elected officials.

3.2.1. *Evry, the French cradle of human genomics*

In addition to performance indicators, such as dashboards and the expansion of the site, we observe a narrative shaped by both institutional communication and the discourse of political leaders. In the manner of Francesco Panese's *Human Brain Project* investigation, it is therefore a case of:

> Tracing the course of the main "discursive events" […] this journey will allow us to decipher the coproduction of scientific credibility and political legitimacy of a grand project shaped, between science and politics, as a prophecy based on a shared imaginary of improving the present (Panese 2017, p. 168).

The legitimacy of the governance of emerging technosciences, in order to mobilize financial, human and political resources, requires the creation of a great fiction (Callon *et al.* 2010), in which the new technology is necessary to solve the major problems facing humanity (world hunger, pandemics, the environmental crisis, the depletion of fossil fuels, etc.)[1]. The fiction of the cluster is based on a narrative of local history, which emphasizes the pioneering spirit of the creation *ex nihilo of* a real campus in a new city where everything was to be built. By way of example, at an event in November 2016, dubbed "Evry, cradle of human genomics", Manuel Valls, former Mayor of Evry and Prime Minister at the time, declared:

> It is in this rather unlikely place that this adventure was born and, basically, scientific research – with its dogmas and its organization, particularly in France – saw the birth of something that did not quite resemble research and, in particular, the alliance between the public and the private sector, but we had to

1 In particular, Pierre-Benoît Joly has highlighted how GMO advocates have had to justify that there is no alternative worldwide to this technology to combat hunger when it is really a question of distribution, not availability (Joly 2015, p. 38).

> fight. Scientific research is projects like this one. Research is also the women and men who, day after day, with this ongoing faith, experiment, compare results, publish, advance our knowledge and thus reduce diseases (extract from Manuel Valls' speech, at the Génocentre, November 4, 2016).

Within the fiction, the problems to be solved and the proposed solutions are formulated without the need to return to the identification of these concerns (Callon *et al.* 2010). Here, the "alliance between the public and the private" shortens a narrative that draws on the historical past of the territory where the biocluster is located. The French Muscular Dystrophy Association (AFM) and its laboratory, Généthon are regularly mentioned in speeches and documents (as evidenced by the historical anchor in the 2001 dashboards, *Genopole, la cité du gène et des biotechs* (Genopole, the gene and biotech city))[2]. The cluster then follows in the footsteps of the AFM, highlighting in particular its historical affiliation with a scientific achievement resulting from Généthon: the first mapping of the human genome. Moreover, Valls uses the lexical field specific to innovation by evoking adventure, scientific discovery and the search for a better world (Sainsaulieu and Saint-Martin 2017, p. 12). We will return to this final point later. Within this lexical field, much emphasis is given to "scientific heroes" (Felt 1993; Callon *et al.* 2010). Indeed, the November 4, 2016 symposium organized by Genopole specifically aimed at paying tribute to the former president of the AFM, Bernard Barataud, and to the "pioneering genomics researchers, Jean Weissenbach, Charles Auffray and Daniel Cohen", evoking "the richness of their research which has made Evry the cradle of French genomics"[3]. Moreover, as we can see from the words of Manuel Valls, political discourse highlights the city of Evry as "this somewhat unlikely place":

> Most accounts of the birth of Genopole on the Evry site, whether journalistic, from the main players involved in its creation, or from sociological analyses, emphasize the metaphor of the Evry area as a "desert". The history of Genopole Evry appears to be an accident, or even "aberration", because there is no *a priori* link between the territorial resources that this geographical area has to offer and the choice made by the

2 https://www.genopole.fr/IMG/pdf/plaquette_genopole_fr_020531.pdf.

3 Genopole's youtube channel: https://www.youtube.com/watch?v=ZejcVGzSEy8.

> public authorities to locate the bulk of the national genomics policy effort there (Peerbaye 2004, p. 156).

This "desert" metaphor contributes to the construction of Evry image as an outsider, a territory far from Paris and not very attractive economically, which has succeeded in gambling on attracting and developing a significant number of laboratories and companies, where most observers predicted failure. In addition to this idea of an impossible, but nevertheless challenging gamble[4], the biocluster project bases part of its legitimacy on the objective of reindustrializing the territory.

3.2.2. *Industrial renewal: the cluster as a solution to local economic development*

The interviews and observations conducted with the biocluster's management, as well as archival documents, highlight the cluster's role as a remedy for the departure of companies, particularly computer companies in the 1990s (Digital, Compaq, IBM). The departure of these companies would moreover have resulted in land being made available for the installation of the cluster:

> I was very aware that Evry, which was a new town, had benefited from resources to attract industrialists and that all these means had disappeared, and even today we see the same movement being perpetuated. The global management of Carrefour has moved to Massy. Crédit Agricole, which has 600 employees, is leaving Massy. It is clear that Evry was not the place… Well, it was artificial […]. All the big crates that had arrived, which is how we found buildings easily, were leaving one after the other: IBM, Hewlett Packard, Compaq, all these people disappeared. The only prospect left was Safran, and Genopole arrived in this context (interview with biocluster management, April 2017).

Genopole is a godsend for the Evry region, at a time when the cluster model is being promoted in all global and European discourse as

4 This rhetoric also flooded the media on the occasion of Pierre Tambourin's departure at the beginning of 2017: the press articles recounting his career and based on his words highlight Evry as both a virgin zone and a minefield for the development of biotechnologies.

the solution to economic competitiveness. In this context, it becomes an asset in terms of governance. Indeed, we can see that the rhetoric developed around clusters is their strength, as it persuades industrial and political-administrative spheres of the validity of the concept. Public decision-makers then have a model to implement to revive the economy. An interview with the management of the biocluster confirms this hypothesis:

As for the local politicians, Manuel Valls wasn't there yet, but for Guyard, the people from the department and even Serge Dassault, who was already there, the idea of a major project like that, without knowing whether or not it would work, appeared to them to be a major element for the problems of image, employment, etc. All these people, from both the right and the left, got behind the Genopole project, and this remained true for twenty years [...] because they had a big project to get their teeth into, so the voters, all that... (interview with biocluster management, April 2017).

Indeed, throughout the survey, regular visits and speeches by public officials confirmed this political enthusiasm for Genopole and, more generally, for the cluster concept. In the deindustrialized area of Evry, the development of the cluster is essential for the local economy, as well as for developing the image of the city as a suburb (Sedel 2009). In addition to the biocluster being necessary for local development, its existence is strongly associated with the discourse of global competition around innovation. From then on, the rhetoric insists on the urgency of aligning with North American clusters.

3.2.3. *The rhetoric of technological backwardness: overcoming French scientific slowness*

The legitimacy of the cluster model as a tool for adjusting between academic science and industry is also based on a rhetoric associated with the diagnosis of global competition and France's difficulty in holding its position in the race (Callon *et al.* 2010). In a study on the media treatment of superconductivity between the United States and Europe, in addition to references to scientific heroes, Felt notes, on the European side, statements that alert to the delay in this technology. He concludes that, in a context of fierce competition, "selling" science has become an obligation (Felt 1993, p. 390). Thus, just as much of the promise and contemporary technoscience is based on the success stories that are currently shaping the economy and society (Loeve 2017, p. 91), Genopole is also being compared to successful cluster models. At a 2014 symposium on the theme of innovation, the then Minister of Higher Education, Geneviève Fioraso, took the opportunity to compare Genopole with the California valley and keep Genopole in the global race:

> All over the world, we find this model of innovation based on large campuses, combining international universities of all sizes, research institutes and start-ups, mid-sized or large companies. This is true for Silicon Valley, where recently I saw a big company, Google, that was founded a few years earlier by students, in dialogue with a researcher in basic biology research, and it was fluid, there were no silos and no one felt threatened by the other. I think it is this type of partnership and dialogue that we need to develop in our country, which is still somewhat compartmentalized. This explains why we are lagging behind in terms of innovation, because I spoke of being in sixth place for research, but we are only in twentieth place in terms of innovation and we are considered to be more a follower than a leader, even as an intermediate country. We need to move up in terms of innovation, even if we have to put rankings into perspective, we still need to know how to tackle problems and how to respond to them (extract from Geneviève Fioraso's speech, at the Genocentre, October 2, 2014).

To make up for this delay, the decompartmentalization of science and industry is considered insufficient. Geneviève Fioraso's speech is a good illustration of the idea of change for change's sake, which is the prototype of the "luring imaginary" (Enriquez 2007, cited in de Gaulejac and Hanique 2015). This imaginary consists of devaluing the current situation in order to highlight a possible future that is necessarily positive. In this framework, resistance to change appears necessarily negative and justifies the race ahead (*ibid.*, p. 21). Thus, "arriving first" and "not lagging behind" are expressions that are frequently used and that insist on maintaining leadership (Bensaude-Vincent 2015, p. 63); as Geneviève Fioraso says, "we are only in twentieth place [...] we have to move up in terms of innovation". This type of discourse is also due to the emergence of figures among political leaders who, because of their background, are sensitive to the rapprochement between science and industry. Geneviève Fioraso, for example, was an executive in a start-up that was spun off from the CEA and then research deputy for the city of Grenoble (Vergès 2014, p. 245).

This rhetoric of emancipation from sclerotic structures[5] is part of a regime of technoscientific promises that must be legitimate and credible. In the first instance, problematization aims to provide ready-made answers and solutions to a problem that has already been identified. Joly points out that the more serious and urgent the problem, the more legitimate the promise of improvement. He uses the example of the urgency and importance of maintaining a position in global competition (Joly 2015, p. 38). From then on, any form of resistance to this narrative appears as resistance to modernity and common sense. Thus, the biocluster's communication endeavors to put into words and images a model that is difficult to contest. Its unquestionable character is strengthened by promises that announce a bright future in terms of progress and innovation.

3.3. Promises of innovation and employment at the territorial level

The synergies made possible by geographic clustering hold promises on two levels: innovation and job creation at the local level.

3.3.1. *The promise of the biocluster: a sustainable environment and the medicine of the future*[6]

The context of economic competitiveness, as we have just seen, legitimizes the action of states, which make a selection in the funding of research by subsidizing "sectors deemed to be promising, priorities, such as biotechnologies, materials sciences, nanotechnologies or neurosciences" (Audetat 2015b, p. 12). These disciplines are emerging, not only because they are deemed to bring about innovation, but also because they proclaim

5 Expression borrowed from Arnaud Saint-Martin during his intervention "Duality, plasticity, duplicity: contemporary uses of satellite observation techniques" at the Journées scientifiques du center Pierre Naville, *Sciences et émancipation. Ce que font les sciences à la société*, December 7, 2017.

6 In a communication brochure posted on March 2, 2018 on the Genopole website, the two strategic axes are "a sustainable environment" and "the medicine of the future". Available at: https://www.genopole.fr/IMG/pdf/2018_genopole_plaquette.pdf (p. 5).

themselves capable of responding to the challenges facing the 21st century, such as global warming, world hunger, disease, etc. (Joly 2015, p. 39). Indeed, the new public management, by generalizing calls for projects as conditions for research funding, mainly through the ANR, has led researchers to deploy arguments in order to obtain these funds (*ibid.*, p. 7). These arguments have a strong tendency to construct scientific promises in order to win the support of financiers, whether private (venture capitalists) or public (ANR, competitiveness clusters, EU, etc.). The growth in the production of these promises does not only concern researchers, but seems to be a phenomenon that affects innovation actors in the broadest sense[7].

Following the lead of laboratory communication departments, which are seeing an increase in staff (Audetat 2015a, p. 76), clusters are also strengthening their communication around emerging science and technology regarding their ability to transform the world and solve its biggest problems. In 2017, the Genopole structure provides 9% of its budget dedicated to communication, which represents a considerable package of €841,000[8]. In a context of technoscientific capitalism, the promise becomes a specific mode of communication (Sunder-Rajan 2006) and, as we have just seen from the creation of the cluster, this department devotes a large part of its strategy to a "promising communication" that consists of elaborating prospective narratives (Quet 2012). Also, in this context, the communication department relies on the emergence of organizations, already mentioned above, such as progressive institutes, communication agencies, etc., specialized in the production and commercialization of knowledge oriented towards the future (Pollock and Williams 2010, p. 525). Indeed, a closer look at the press release for the "*Vivre l'innovation aujourd'hui et demain*" (Living Innovation Today and Tomorrow) conference to celebrate the biocluster's 20th anniversary shows that it was written by both Genopole's institutional communication department and the firm Andrew Lloyd & Associates, which provides strategic communication services to "leading companies and

---

7 Several works in sociology and economics emphasize the weight of expectations regarding research and development investment in the formulation of promises (Rosenberg 1983; Dosi 1988). Marc Audetat insists on the development of a sociology of expectations (Audetat 2015) or *sociology of expectations* (Brown 2003; Borup *et al.* 2006) that focuses on the way in which the research funding system gives rise to predictions that rely on imaginaries that come to justify the promises made (Joly *et al.* 2010; Jasanoff and Kim 2015).

8 Figures presented at the PIG staff meeting on February 28, 2017.

organizations involved in a wide range of emerging technologies"[9], including biotechnology, pharmaceuticals, electronics, photonics and aeronautics.

As early as 2003, the communications department developed a storytelling approach (Quet 2012) with the publication of a book devoted to the five years of the cluster *Il était une fois Genopole* (Once Upon a Time There Was Genopole)[10]. The chapter entitled "*Un avenir prometteur*" (A promising future) predicted, even then, that the cluster would be "inventing the medicine of tomorrow" (*ibid.*, p. 28), thanks in particular to the establishment of the Centre Hospitalier Sud Francilien (South Paris Hospital Center) near the campus, following the merger of the Evry and Corbeil-Essonnes hospitals. This move promises that patients and their families will be able to rapidly benefit from discoveries and innovations due to the proximity of the cluster's companies and laboratories. In 2003, Genopole's communications, strongly influenced by the identity of the French Muscular Dystrophy Association (AFM), were also part of a unifying belief in the fight against disease. The first dashboards, communication documents and the website, are built around the mantra "Succeeding together in biotechnology" which extends the foundations laid down by the AFM. Recently, it has been replaced by the fashionable term "innovation", with the slogan "*Vivre l'innovation. Innovation today, every day*". In the activity report published in 2016, the various themes presenting Genopole's activities are called "resource creator", "project federator", "idea developer", "technology cultivator", etc. These names are taken from the vocabulary of innovation already mentioned: adventure, scientific discovery, the search for a better world, etc. (Sainsaulieu and Saint-Martin 2017, p. 12). Beyond this lexicon, the promise made by the biocluster in supporting biotech companies and life science laboratories (many of which conduct health research) is to offer a sustainable environment and the medicine of the future. The 2016 report thus states, "Genopole contributes to hatching the innovations that will improve our health, living conditions and environment" (p. 3). Alongside this prediction that society will be better off as a result of the innovations produced by the cluster, there is a promise of job creation. Indeed, by 2025, the biocluster plans to "become the capital of genobiomedicine".

---

9 http://www.ala.com/clients/.

10 The chapters of the book can be downloaded at: http://www.genopole.fr/Le-Livre-anniversaire-des-5-ans-de.html#. WNEKuxI19EI.

3.3.2. *Becoming the capital of genobiomedicine, creating jobs for innovators*

In the same way that the European institutions have set up the, as its name suggests, future-oriented research and innovation program "Horizon H2020", the biocluster has also set up a forward-looking strategic plan called "Genopole 2025". In 2011, Thierry Mandon, then President of Genopole, recalled the context in which this strategy was developed. Having discussed once again the successful challenge of converting the region and the biocluster's national reputation, he insisted on positioning the cluster at a global level through the Genopole 2025 strategic plan in order to generate the confidence of public financiers:

> We have therefore launched a study on the future of Genopole up to 2025, with a view to it becoming a major player on the world stage. This reflection has focused on strategic directions and on the new mandate we need to seek from our partners: the state, the region, the department, the Evry agglomeration community, the major research organizations and all our partners from day one. By drawing up this Genopole 2025 strategic plan, we will be able to renegotiate a new commitment with the public authorities up to 2012 and to answer certain questions about resources […]. On this basis, we will be able to enter into a new contract with our financiers and encourage them to understand that the transformation of Genopole into a world-class cluster requires a concentrated effort. France invested, but was quickly overtaken by the United States, Germany and England. It must therefore reflect on its ambitions by 2025 and concentrate the related resources (extract from the speech by Thierry Mandon, president of the biocluster, at the General Assembly of the cluster's leaders and decision-makers, Genoforum 2011, December 16, 2011).

This extract from the speech accurately reflects the funding mechanism highlighted by Van Lente and Rip under the term "strategic science" (Rip and Van Lente 1998). The rhetorical space used by the biocluster produces promises aimed at capturing resources on a competitive basis with respect to financiers: "to encourage them to understand" the "concentration of efforts" in the face of "the United States, Germany and England"[11]. The dominant promise is to make Evry "the capital of genobiomedicine"[12]. This is a neologism commonly and almost exclusively used[13] by Genopole to refer to the future challenges of medicine using a genome analysis approach. To achieve this, the cluster has broken down its Genopole 2025 plan into eight objectives[14]. In addition to promises relating to the development of new technologies, working methods, land acquisition, etc., the biocluster's vision is based on a figure who embodies the harmonious combination of science and industry, the entrepreneurial researcher or heroic innovator (Callon *et al.* 2010), with both scientific and entrepreneurial qualities and always in search of innovation[15]. This anticipatory account of a generation of innovators in the making is clearly evident in the words of the French Education and Research Minister during her visit to Genopole:

> We must know how to make the most of our assets in order to build together this model of innovation that creates jobs and progress [...] and I will conclude by quoting Seneca: "It is not because things are difficult that we do not dare; it is because we do not dare that they are difficult." This is how we must act now in France, to take initiatives, to transmit the taste for risk to young people, because if we no longer have this taste for innovation, if we hide behind the precautionary principle, which is useful but must not become a principle of inaction, then we are not proposing a future in which young people can project themselves and we cannot be surprised that they go to countries that, in turn, become more pioneering. Let us find this spirit of adventure, this spirit of risk, this taste for daring, this is how we will build a world that is fairer, more creative and a provider of more jobs and progress. Thank you for contributing to this

11 Not to mention the subnational competition.

12 media.genopole.fr/medias/domain1/media113/18142-v7m2uij2nl.docx.

13 A quick search on the Internet shows that the term only refers to web pages dedicated to Genopole.

14 The eight objectives of Genopole 2025 can be found at: https://www.genopole.fr/Notre-strategie-a-l-horizon-2025.html#. WqejfBPOWi4.

15 Genopole has been honoring these innovators even more since 2011, with the creation of the *Genopole Young Biotech Award*, which rewards the young start-up entrepreneurs who are supported by the cluster.

> (extract from Geneviève Fioraso's speech at the Génocentre, October 2, 2014).

This heroic and adventurous spirit, in addition to being promised, is even invoked: "we must know", "this is how we must act", "let us find out". However, above all, it illustrates a dimension developed by the sociology of expectations that links promises to the calculations of public investors with the concept of promise requirement cycles (Van Lente 1993). These "cycles of promises and requirements" show how promises are put forward and then how they, in turn, create demands, even requirements.

The cluster then uses the promise of employment in a logic of resource capture that it regularly recalls. For example, the dashboard released in 2017 promises the following future:

> Genopole aims to develop a world-class biocluster on the Evry-Corbeil site and to become the European capital of genobiomedicine and bioindustries by 2025. Its objective is to bring together 30 to 35 research laboratories and 130 companies, representing 4,000 to 5,000 people (2016 biocluster activity report, *Vivre l'innovation. Innovation today, every day*, p. 14).

For nearly seven years, communication documents have predicted these numbers. However, a closer look at the statistics, presented by the cluster a few pages later, reveals the difficulty in keeping the promise.

Between 2005 and 2016, the cluster's workforce grew from 1,827 people to 2,475, an increase of 628 people in 10 years. Over the same period, between 2016 and 2025, the biocluster promises to double its workforce, which would represent an increase of 2,500 people. There is therefore a clear gap between the actual increase in staff and what the institutional communication promises, in the same way as some observers have noted the growing gap between promises and actual innovation in the pharmaceutical and nanotechnology fields (Audetat 2015a). In this case, the cluster has failed over the years to fulfill its commitment in terms of jobs. Given these limitations, how can we explain the fact that the developed narrative is not running out of steam?

3.3.3. *The naturalization of the cluster effect: an unquestionable concept?*

Questioning the nature of the narrative developed by the cluster leads us to question the sustainability of the proposed imaginary. On the economic side, Frédéric Lordon developed the principle of the strength of simple ideas in economics (Lordon 2000, pp. 184–187). According to Lordon, these simple ideas can take the form of a watchword, synthesizing the overall project (e.g. "stability"). In the same way, innovation is a watchword, while it is reified and naturalized, without being questioned since it is posed upstream of reflections (Sainsaulieu and Saint-Martin 2017, p. 17). Indeed, the promise is embodied in a self-fulfilling prophecy that naturalizes technical change (Callon *et al.* 2010). This naturalization takes place in common places such that "if the necessary resources are gathered, the technique will ineluctably bring solutions", or "if we do not invest in this technology, others will do so before us". It is then difficult to conceive alternatives to this kind of prognosis.

A critic, like Lordon, of the use of the lexical field of ease, Granger believes that the knowledge economy has "the simplicity of beautiful theoretical nesting" (Granger 2015, p. 65). He then takes up the criticisms regularly directed towards higher education, as noted by Christian Laval in *Le nouvel ordre éducatif mondial* (The new global educational order), again beginning with the conditional conjunction "if": "if there is unemployment in France, it is because university training is not adapted to the needs of business", or "if the economy is at half-mast, it is because research is too academic, too fundamental and barely oriented towards bringing innovative products or concepts to market" (Dardot and Weber 2002). These cause-and-effect relationships (if such and such a situation either does or does not exist, we must expect or rely on such and such a situation) call upon laws that conceptualize their naturalization: Moore's law, Gabor's law, etc. (Callon *et al.* 2010). Moore's law, in particular, appears to be a matrix of technoscientific promises, as a metaphor for technical progress in general (Loeve 2017). Indeed, this law derives from a prediction by Gordon Moore, a computer scientist and co-founder of the Intel corporation, who predicted that the number of transistors, the main components of processors, would double every two years.

In his article "La loi de Moore: enquête critique sur l'économie d'une promesse" (Moore's Law: critical analysis of the economics of a promise), Sacha Loeve shows how this law is a prototype of a technological roadmap that brings together a diversity of actors towards a common cause. Its naturalization is such that few people have been interested in its construction and validity, the only questions being of the "how to keep it going?" type. By shedding light on the fabric of this law, Loeve shows how this prophecy was constructed after the fact. To Loeve, it is not an industrial lie but an invention of the present that modifies the meaning of the past, which creates "the possible behind itself" (*ibid.* p. 110) and thus turns into a perpetual promise. Also, similar to the common places highlighted by Dardot, the credibility of the cluster is intimately linked to the simplistic naturalization that results from it. In his case, the prophecy is as follows: if companies, laboratories and training are grouped together geographically, the synergies between researchers and industrialists will be strengthened and will lead to innovation. This principle would then constitute what Genopole managers have referred to throughout the survey as the "cluster effect"[16], evading the idea that this law can be disproved in reality:

> But because I believe in the cluster effect [...]. For me, a cluster is a place where you can go to a conference without difficulty and exchange with other people without problems. That's why I insist on a sufficient critical mass [...] so the idea is to reach what I call an irreversibility threshold of people on the site. In Sophia Antipolis, no one notices when a company leaves, so the idea is to get to that point as quickly as possible, because it's something I fear, to see the funds granted to the PIG for its development. For me, it is essential to reach these decisive steps that I estimate at 4,000/5,000 jobs, roughly 150 companies [...]. I remember the Christian Blanc report, which really hit me. Michael Porter, I already knew... I had the cluster effect in my mind as far back as 2004. And above all, I had to distinguish between the Genopole cluster and the Medicen competitiveness pole, which I was partly in charge of at the time, by saying that a cluster is more restricted, whereas a competitiveness pole is

16 "Genopole creates the right conditions for meetings between the site's players, companies and laboratories, which is a necessary condition for creating the biocluster effect." Genopole-cree-les-conditions-favorables-aux.html#.WqpNbRPOWi4.

> more diffuse, more global. They're not on the same scale! (interview with biocluster management, April 2017).

On the one hand, there is a fairly advanced conceptualization in these remarks, "cluster effect", "critical mass", "irreversibility threshold", and, on the other hand, two reference figures: on one side, the figure of the North American expert theorist, Michael Porter, and the political figure at the origin of the assimilation of the concept in French public policies, Christian Blanc. The law described here is based on the idea of entrainment found in economics on the network effect in certain technologies, such as communication services, which considers that the technique passes through network exchanges that represent the source of the mechanism that causes this type of effect[17]. The cluster effect is then based on the economic idea of positive externalities: the mutualization of equipment, the grouping of skills, the intensification of exchanges, etc. The last of these is the black spot of many clusters. While most have succeeded in bringing together scientific and industrial skills and cutting-edge infrastructures, few of them spontaneously see the interactions developing within them and stick to the benefits in terms of marketing, brought about by accreditation practices.

Ultimately, belief in the promise of the cluster effect is not actually oriented towards a desirable future, but serves more to pull the present along (Bensaude-Vincent 2015, p. 63). This promise embodies more of a "rational mobilizing myth" (Hatchuel 1996) that allows for the coordination and federating of actions (the national program on precision medicine, for example), the setting up of infrastructures such as shared platforms, the mobilization of researchers in business creation processes, etc. Thus reified and naturalized, it becomes difficult to question it. Indeed, throughout the immersion survey within the cluster, the simple idea of cluster effect, in Lordon's sense, was so strong that it was never questioned. The common belief, shared by the different layers of managers, was that if the cluster effect is not working optimally, it is because Genopole has not yet found the modalities for its application in its territory. Indeed, in the second part of this book, we will see how the cluster has been trying, since its creation, to develop interaction between science and industry in an iterative process, through a system of networking.

---

17 For example, the value placed by a consumer on a network service increases as the number of consumers of that service increases: https://www.universalis.fr/encyclopedie/externalite-economie/#i_55668.

PART 2

Prevented Synergies: the Case of a Biotechnology Cluster

4

Networking Systems: Repeated but Hindered Initiatives

As soon as clusters begin to bring together more and more members, businesses or laboratories, their administrators think about ways to encourage synergies between them: seminars, workshops, calls for projects (for competitiveness clusters), afterworks, etc. Several formulas are tested so that entrepreneurs and researchers can meet and then develop collaborations. In the cluster studied, the networking system adopted follows much the same approach. It consists of bringing together workers from the site's companies and laboratories around common themes or activities, outside working hours (lunch breaks are usually preferred) on a voluntary basis, and then waiting for spontaneous cooperation to emerge. In the example of Genopole, the system is deployed in three main ways: a policy of scientific administration through the organization of seminars and workshops, the pooling of equipment intended to be shared and the setting up of convivial and sporting activities.

4.1. Scientific and industrial administration: establishing a recurrent event

Although the scientific and industrial administration of the cluster has been a common theme throughout the history of the biocluster since its inception, it was only in 2012 that a real administration policy was implemented, including the recruitment of a person dedicated to its deployment.

4.1.1. *The emergence of an intermediary figure: the cluster administrator*

In 2011, a general meeting brought together the managements of accredited companies and laboratories to discuss the themes of "Company-Laboratory Collaboration", "Biopark Administration" and "Site Life". The names of these three workshops showed that the emphasis was on the cooperative aspect. To achieve this, it was necessary to "develop the scientific cluster into a real biocluster"[1]. The idea of transforming a biopark, with its geographical grouping of companies, research laboratories and training courses in the biotechnology sector, into a biocluster with individual and institutional relationships between these three players, is present in management literature (Hamdouch 2007). Up to this point, the issue of administration had been handled either by the Research Department, the Communication Department, or both. Although meetings had been set up, scientific administration had not been the subject of a policy on its own, but rather part of more general projects attached to other departments.

From 2012, the biocluster intended to entrust the development and implementation of an administration policy to a dedicated person within the Communication Department. Indeed, the management expected this position to define a strategic action plan for the administration of the biocluster, the objective of which being to develop meetings, interactions and collaborations between (1) the actors of the site and (2) with external partners, as well as (3) the attractiveness and visibility of the biocluster. Therefore, through this recruitment, the biocluster expressed its desire to develop an administration strategy. The role of a person in charge of animation was therefore part of the biocluster's intention to systematize networking that had, until this point, been restricted mainly to the organization of scientific seminars.

Note that this type of position is not unique and can be found in other clusters in France: non-exhaustively, these include Silver Valley in Île-de-France[2], the SPN cluster *Professionnels du numérique dans le Poitou*[3] or

1 Transcript of the General Meeting of Cluster Leaders and Decision Makers, Genoforum, December 16, 2011, p. 4.

2 http://www.silvervalley.fr/Animation-du-reseau.

3 http://www.spn.asso.fr/wp-content/uploads/2015/08/Offre-Animateur.pdf.

even the *Rhône-Alpes éco-énergies*[4] cluster. In order to respond to this professionalization, in 2014, the University of Strasbourg even delivered a Master 2 training course entitled "Cluster and Territorial Network Administrator", the educational objective being:

> To train cluster and trilingual territorial network administrators to develop innovative projects by creating synergies between businesses, universities, research centers and public administrations. This means: detecting the driving forces (public-private territorial actors), mobilizing them around projects based on individual needs and creating an administration and institutional cohesion unit for long-term action[5].

A study of the courses offered online shows that this is the only training of its kind in France for the moment. Nevertheless, the very existence of training for such a specialized type of position testifies to the importance that these roles as intermediary between "companies, universities, research centers and public administrations" (see extract above) are likely to take on.

The job description circulated by the biocluster called for the establishment of internal "bridges" between the Enterprise department (*Genopole Entreprises*) and the Research and Platforms department (*Genopole Recherche et Plateformes*) at Genopole. In other words, the Animation Manager's responsibility was to create the conditions for internal collaboration between independent departments serving different interests: laboratory support and scientific equipment development, on the one hand, and business development, on the other hand. Ultimately, the main task of this position was to bring together companies and laboratories at the biocluster level, based on the creation of a cross-functional project team combiningexecutives from Genopole Entreprises and Genopole Recherche. Since 2014, the team has been made up of two business managers and the head of business development at Genopole Entreprises, a project manager and the head of platforms at Genopole Recherche, and the director of

4 http://www.aradel.asso.fr/upload/File/offreEmploi/fiche%20de%20poste%20cm%20animati on_405142.pdf.

5 https://itiri.unistra.fr/uploads/media/FicheAnimateur_de_cluster1.pdf.

Genopole Communication[6]. The person recruited for the position of Animation Manager embodies a combination of the two dimensions of public research and the business world. She holds a PhD in biology and has subsequently worked in marketing, notably in the pharmaceutical industry. This type of profile is also prevalent at Genopole Entreprises: several project managers originally had a scientific background, which they supplemented either with experience in the industrial sector or with a Master of Business Administration (MBA) in commerce, industrial property or finance. The Genopole Recherche et Plateformes project managers also have scientific backgrounds, including theses in biology. They have spent their careers in laboratories and then in scientific administration at institutes such as the National Institute of Health and Medical Research (*Institut National de la Santé Et de la Recherche Médicale*), or the National Institute of Agronomic Research (*Institut National de la Recherche Agronomique*), etc. As a result, the most notable feature of the Animation Manager and the Genopole Entreprises teams are their profiles of intermediation between science and industry. The Animation Manager works to create cooperation between researchers and industry, while the Genopole Entreprises team assists and advises researchers in business creation.

As a social intermediary, the Animation Manager organizes events that bring together staff from Genopole labs and companies: on the one hand, regular events, such as welcome sessions (WS) for newly accredited structures, themed workshops (around specific topics, such as intellectual property, patent law, etc.) and science and technology seminars; on the other hand, the formalization of working communities around a specific field of research and activity (immunology, synthetic biology, etc.) or common issues (such as management committees and PhD student clubs).

4.1.2. *Networking and renewing acquaintances among cluster members through regular events*

The WS, held around three times a year, involves welcoming companies that have recently been accredited. During a morning session, these companies present their activities to an audience composed mainly of

6 Within the framework of the CIFRE, I was also part of this team as a research officer attached to the General Management.

executives or managers working in companies already established[7]. While the survey revealed a lack of awareness of the full range of activities, the WS were the events most spontaneously mentioned, whether in informal exchanges, interviews or data from the questionnaire (see Figure 4.1). The concept is appreciated and considered useful for getting to know new members, but difficulties in terms of time availability are often mentioned:

> I use the administrative services when I can… The days where the companies present themselves, I found those very interesting. I went once, but after that I've rarely been able to go. It's a great idea, but it's a matter of availability afterwards… (interview with Blandine, director of an accredited laboratory, October 2015).

This availability is even more reduced during other events, such as workshops around a specific theme (patents, preparing clinical studies, recruitment) and science and technology seminars of a more academic nature. Like the WS, these events are more relevant to business and laboratory management, rather than to all employees: 12% participated in the seminars and 9% in the workshops (see Figure 4.1), whereas the questionnaire for managers shows that 40% of them participated. This difference in attendance between managers and their employees is expressed in the words of some respondents:

> Yes, there are indeed workshops. I did one, but since then I haven't been interested in any, because it's more focused on managers with breakfasts, *biz dev* [business development], how to develop internationally, etc. […]. For me, there should be science-oriented themes, like some kind of congress, small conferences on a theme, each company doing a conference every week, and it rotating, so that everyone knows what the others are doing (interview with Leïla, director of studies at an accredited company, July 2015).

---

7 Indeed, among the people who answered that they had already participated in the WS in the questionnaire distributed online, three quarters (76%) work in a company, and are mainly managers, or even company directors. As the questionnaire was distributed on an individual scale (responses from individuals regardless of their status) and on an organizational scale (responses from managers), the results indicate that 18% of the individuals surveyed had already participated in a WS, compared to 56% of the managers surveyed.

In 2016, there were 30 face-to-face events in total, equivalent to three events per month if we exclude the summer period, which is generally quieter. The number and frequency of events thus seek to create a recurrence of meetings between cluster members, so that they can initially meet and then recognize each other later.

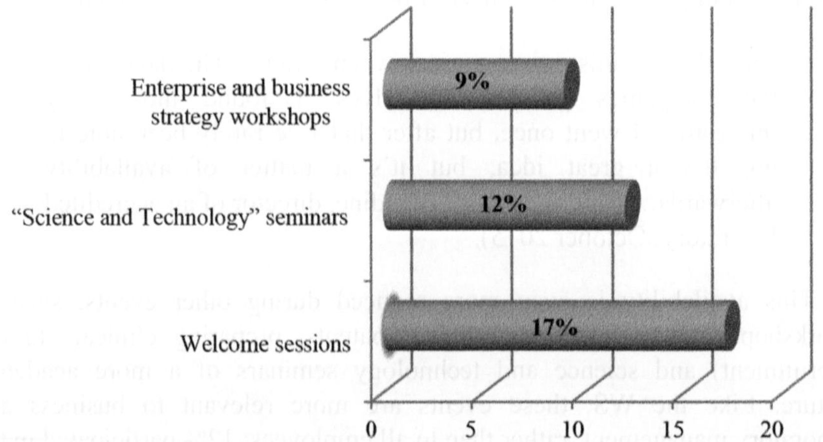

Figure 4.1. *Cluster workers' participation in professional events (source: graph constructed from questionnaire data sent to 102 people working in the cluster between February and April 2016). For a color version of this figure, see www.iste.co.uk/vallier/clusters.zip*

On a regional and national scale, the biocluster organizes conferences open to outside audiences (industrialists, researchers, institutions, etc.) that are likely to develop into areas where interaction and socializing can take place. These events generally take place in a venue located on the AFM premises: the Génocentre. It is the "conference and convention center dedicated to science and business"[8] and can accommodate up to 700 people. The biocluster generally organizes at least one conference at the Génocentre per year.

In terms of attendance, according to the questionnaire survey, 56% of managers interviewed and 27% of individuals of all statuses attended the conferences. The various observations and participations in these events suggest that they are more tools for promoting the site than vectors of

8 http://www.essonne-parisud.com/lieux-de-seminaires-sans-hebergement/fiche-tourisme/genocentre/.

cooperation. Two types of colloquia should be distinguished. First, scientific conferences: in particular, the International Innovation Conference in 2014[9], the Personalized Medicine Conference in 2015[10] and the Synthetic Biology Conference in 2016[11], which contributed to the cluster's reputation at national and international levels. In addition, the conferences were more focused on the cluster's accredited members, such as the Anniversary conferences, organized for the cluster's 15th anniversary in January 2014 and for its 20th anniversary at the end of 2018, as well as the Genoforum conferences in 2011 and 2016. If we take the latter as an example, a preparation document stated the principle of the event:

> As a showcase for Genopolitan know-how for the biocluster's scientific and industrial community and the Ile-de-France region, Genoforum is an opportunity to reconcile science, business and conviviality via: a presentation of the strategy developed by the Genopole PIG within the biocluster's perimeter, plus the opportunity to bring together the players who make the campus news (Genoforum preparatory note, February 5, 2016).

Indeed, while their formulation indicates moments of internal exchanges within the cluster (anniversary, forum), these events are seen as "showcases" that contribute to the reputation of the biocluster. Ultimately, this type of event, which is part of the cluster's administrative policy, seems to be more of a tool for promoting the site than an opportunity to highlight and encourage synergies.

4.1.3. *Fostering communities of practice: the creation of thematic bodies*

In addition to this protean event strategy, Genopole aims to encourage formal and informal groups of individuals with common interests or research themes. Three mechanisms were observed during the survey which demonstrated the application of this principle: a management committee, an immunology theme group and a PhD students' club. In this logic, the sharing

9 http://www.genopole.fr/spip.php?page=rubrique_event&id_rubrique=784&event=784#. WSRdHhPyjHg.

10 http://www.genopole.fr/Videos-du-colloque-medecine-personnalisee.html#. WSRcNBPyjHg.

11 http://www.genopole.fr/-Synthetic-Biology-From-Ideas-to-Market-.html.

of an interest or a common problem would lead the actors to build together a shared understanding of the activity. The constitution of these groups echoes the theories developed in knowledge management around the concept of "community of practice", which is defined as:

> A group of individuals who share an interest, set of problems, or passion for a topic and who deepen their knowledge and expertise in that area through ongoing interaction (Habhab-Rave 2010, p. 45).

Wenger distinguishes the community from the group, team or network (Wenger 2005), asserting that what defines community is mutual commitment; it is not simply a set of individuals with common characteristics (*ibid.* p. 83). This is, however, the bias of the biocluster: to bring together individuals with common characteristics, whether in terms of their status (in this case, managers or PhD students) or their research theme. However, mutual commitment is not based on shared attributes, but on the practice of the individuals in the community. From this point of view, the biocluster's science and enterprise steering committee (*Comité d'Orientation Science et Entreprise du Biocluster*, COSEB), that is, the cluster's management committee, is having difficulty forming. One of its members explains these difficulties, one year after its creation in 2013:

> The objective of COSEB is to achieve fruitful exchanges between researchers but also between entrepreneurs, that's what really characterizes COSEB. We try to find a balance between public and private laboratories and entrepreneurs and we try to exchange. So, as the COSEB president would say, "we've already got to meet and talk to each other, to speak the same language", which is no mean feat… But it's been through regular meetings, about every two months, that we've been able to discuss the problems and needs that we have at Genopole. We've tried to create small events, tried to do concrete things, so it's pretty trivial, but to make sure […] that there is a kind of synergy that is a little stronger and that things really emerge from the fact that we are all grouped together in the same location. So it's in line with the objectives of the PIG, but we have our independence. We try to help them to see how to improve things, because it really is based on the concrete, on the fact of being the most consecutive. The risks, however, are

> that actors don't completely see the interest, because we don't really do concrete things that bring something to the entrepreneurs, and we are neglected because of a lack of interest. This is the major risk (interview with Emmanuel, laboratory teacher-researcher and platform manager, July 2014).

In this interview extract, we understand that the more the group meets, the more it creates a common language and a form of mutual trust that leads everyone to express their needs and problems. Nevertheless, while the literature generally concludes that the emergence of a common language and trust between actors leads to strategic exchanges (Estades *et al.* 1996; Galison 1997), we note here that this concretization struggles to emerge.

In 2015, two years after the creation of COSEB, a new modality was tested using "thematic clusters". The idea was still to bring together researchers and industrialists in working groups, but unlike COSEB, where all accredited members were invited, here the idea was to bring them together according to their field of research and activity. The management of the biocluster saw all of its members as "links" that needed to be connected: firstly through training, then research and finally industrial and clinical applications. Nevertheless, these links did not belong to the same value chain, hence the idea of grouping them by theme. The literature on communities of practice emphasizes the need for a common theme, so that different stakeholders understand each other well. They would then be characterized by a shared language or a common subjective view of the world developed through their common experience (Brown and Cook 1999). A similar philosophy was at work with the creation of the first thematic cluster on immunology. However, the needs identified by the group in question did not appear to correspond to the biocluster management's instruction to create partnerships. Indeed, for the members, the two priority objectives were, firstly, to get to know each other better and, secondly, to bring immunology specialists to the Paris Region for conferences on the latest advances or technologies in the field. One of the members of the thematic group on immunology explains how it worked and the different choices they made:

> The idea was that, since the labs' activities may be quite different, we said, "Let's try to make thematic groups". So we made a Genopole immunocluster group to interest immunologists in the broadest sense of the term, both from an

> academic and biotech point of view; so we formed a small committee that mixed biotechs and academic labs. We made a first presentation, there were 20, 25 people on therapeutic antibodies, the next one will be me presenting, so we'll see how it goes… The other idea was to set up a LinkedIn group to share news on the world of immuno, or on everything related to the use of antibodies, or other things. Everyone can contribute to it, but […] they don't really use this service (interview with Alexandre, director of an SME that left the cluster in 2016, September 2015).

Once again, we can see that the partnership dimension is not the priority of the actors involved. They seem to be more concerned with exchanges of information and advice or the support of scientific experts from outside the cluster than with the integration of their respective structures into a value chain.

Finally, the last group observed is part of a strategy to bring together PhD students and postgraduate fellows from the different laboratories of the biocluster. Unlike COSEB and the thematic clusters, which were simply observed, the analysis of this last group is the result of the observation of a participant, both as a salaried PhD student of the biocluster and as a member of the project team supervised by the Animation Manager, who considers that PhD students are a pool of potential future entrepreneurs, likely to settle on the site in the long-term. It seems to her that it would be better to rely on the potential of PhD students, who are already on the site, for the creation of companies, rather than to bring in companies or young entrepreneurs from outside the site. In her opinion, we should base our approach on that of Minatech[12] and the Doctoral principle, which aims to organize meetings between PhD students and businesses. These meetings may take the form of presentations of work to industrialists, teams of PhD students responding to a company's problems over a given period of time, or even regular meetings between students and recruiters. This approach is seen as a way to mobilize PhD students in a fun way, while helping them formalize their scientific work in a way that is understandable to potential investors or employers.

Thus, the first "ApéroDoc" organized on the cluster brought together about 30 PhD students over a drink, divided into several groups, who had to

12 Minatech is a cluster that brings together scientific and industrial players as well as training courses in the Grenoble area in the field of micro- and nanotechnologies.

introduce themselves to each other, and then discuss the actions to be implemented. Throughout this event, the Animation Manager highlighted collaboration projects between them or with companies, or the creation of companies based on their work. The second event, organized six months later, was only attended by four people. Nevertheless, the group got its second wind a few months later thanks to the involvement of a postgraduate student from the site, who started to organize "post-thesis" meetings. The first two of these were on the theme of "What market for PhD employment in France?" led by a consultant from the Health division of the recruitment firm Michael Page, and "Applying for a job after your thesis or post-doc", led by a recruitment officer from Kelly Scientifique[13]. This theme was then eagerly exploited by the PhD student group, who then organized a series of conferences aimed at drawing up portraits of former PhD students on the theme "They defended their thesis, what did they become?" These meetings became an opportunity to shed light on the jobs available to postgraduates.

There is a consistent theme running through the three examples of thematic group formation described above (COSEB, the immunology cluster and the PhD students' club), in that they were all initiated by Genopole and created from scratch as part of its administrative policy. This involvement has consequences in terms of the expectations set out by Genopole, whether that means taking part in the overall strategy of the biocluster and setting up partnerships for COSEB and the thematic clusters, or engaging in business creation for PhD students. In the end, however, the actors involved in the three experiments developed common themes and interests outside the scope of the cooperation between science and industry desired by the cluster. In fact, they were unanimously more interested in the creation of groups from which they could seek advice and information, whether they were managers, seeking to develop a network based on trust, researchers and industrialists specializing in immunology, wanting to join together in the search for information from other specialists in the field, or PhD students, looking for answers to a common concern about their professional opportunities.

4.2. Sharing a technology platform: mutualization or collaboration?

One of the particularities of clusters is the pooling of resources that could otherwise not be paid for by a single company or laboratory. These shared

13 These two events were quite enthusiastically attended, with between 20 and 30 PhD students present at each.

spaces, beyond their instrumental function, are also considered as places conducive to the development of common projects.

4.2.1. *Resources as an intermediary: a policy of sharing expensive equipment*

Based on the literature, clusters can be said to be an intermediation structure. In this sense, intermediation must be understood as a means of compensating for the imperfect circulation of information between suppliers and customers and facilitating contact between two worlds (businesses and laboratories) with heterogeneous knowledge and action logics. The structure would thus be organized on a continuum of action whose two ends are "network" and "actor" intermediation (Branciard 2002). The objective of network intermediation would be the dissemination of information, potential partnerships, obtaining financial aid, etc., while actor intermediation is involved with backing technological and scientific development projects. The cluster, for its part, implements a policy of shared equipment which, in line with this analysis, constitutes a third path that can be described as "resource" intermediation. Thanks to its financial backers, in particular the Regional Council and the Department, the structure effectively offers equipment, mainly installed in laboratories that would not have been able to finance it themselves. In total, the biocluster has 19 shared platforms and a technical platform. For the life sciences, this requirement is based on the use of instrumentation and new information technologies to generate, store, analyze and represent vast quantities of data.

These platforms come in various forms: animal houses and imaging facilities for producing and analyzing transgenic animals, DNA sequencing equipment, advanced genetic methodologies for pathologies, DNA encapsulation, etc. Thus, they form a set of resources available within the perimeter of the cluster. Historically, however, the majority of platforms in France were built around instruments within universities, university hospitals and laboratories attached to public research organizations (INSERM, CEA, INRA, CNRS), in order to develop internal research programs (Mangematin and Peerbaye 2004, p. 705). These platforms then opened up to other users from other laboratories, then from companies or industrial laboratories, thereby creating new markets. Although the biocluster platforms were created for the Genopolitan community, some are more widely used, including commercially, and are the result of initiatives led by research organizations.

Accordingly, the legal mode of execution, the fee schedule and the terms of access are defined by each host structure. That said, the principle of granting preferential rates to French public laboratories seems to be a rule shared by most platforms. This is notably the case for a bioinformatics platform. With its "France Génomique" accreditation, it has benefited from an "investment for the future" to keep France at the forefront of genomic data production and analysis. Its services, therefore, have a national scope. Thus, in order to guarantee quality access, some platforms are developing systems that allow remote use, making geographical proximity incidental. This is particularly the case for the bioinformatics platform, which allows remote work on genomes, whether or not they have already been made public:

> It is a software environment [...] with a certain number of public genomes, there are less than half, about 40% today. These genomes are accessible, there is no need to "log in", we are in "guest" mode. [And with private genomes] when people submit their genomes to us, what they want is for us to do the annotation, There is a coordinator. We interact with one person, that person then manages it. He has a user-friendly interface available and he can then create accounts for all his collaborators wherever they are in the world, to produce analysis and expert annotation (interview with Edith, laboratory CNRS research director, February 2017).

Apart from an optional four-day training course in the bioinformatics laboratory that teaches the prerequisites for using the platform, users can enjoy the platform independently, without physically being in the same place. Consequently, the platform is even able to offer services on an international scale, as shown in the extract above.

Other platforms, on the other hand, have a much more local focus. Although they are theoretically open to all users, they are essentially aimed at the members of the laboratory to which they belong. For example, let us listen to a platform manager, who explains that, ultimately, the majority of users are either directly in the laboratory, or in the start-ups created from the laboratory's research:

> We have many academic researchers in the laboratory who use the platform. And we have an activity that is really on the rise, these are the start-ups that we have aggregated around the laboratory, which are ultimately spin-offs of the laboratories, of the technos [technologies] that were developed in the research teams […] There is, for example, a company that was created and that uses the platform a lot and two other companies that are in the process of being set up and that also use the platform, so that is the biggest use. Then we have partners outside the laboratory, but who are more marginal. We have an academic partner who uses the computing resource, and then, I don't have the details, but we have three or four companies that come on an occasional or regular basis. However, it's very, if you will, light compared to the platform's overall activity (interview with Gwenaël, laboratory research engineer, February 2017).

In this case, the platform is truly at the heart of a strategy of the laboratory and its spin-offs, which all use the same technology, and in the end, it is of little use to the rest of the biocluster's users. If we take a closer look at the use of the platforms by the 102 people who took the questionnaire, we find that seven platforms have never been used by the panel.

On the other hand, some are used more than others, such as the animal house and the imaging and technical platforms. The latter, which is found in the business incubator, differs from the platforms in that it offers only technical resources, whereas the platform offers, in addition to its equipment, skills through the personnel made available. The animal house and the imaging-cytometry platform are essential facilities in the field of life sciences. In addition, their presence has been mentioned, in many informal and informal discussions, as a factor in attracting new companies, as the biocluster's management explains:

> At the start, we only saw resources for better working in the instruments we bought. Then, afterwards, we had this very

> political vision that a good platform, like a good pet shop, can be an attractor of companies and an element of conviviality, exchanges, etc. (interview with biocluster management, April 2017).

In this extract, we note a logic change regarding the platforms, from a positive tool for accredited members to an instrument of attractiveness and exchange. When a director of studies of an accredited company located in the incubator is asked about the reasons for the company's location, she explains that the animal house is the main factor:

> I think it was financially because they looked to locate in Pasteur, in several places, but it was mostly related to the animal house, because it's a very nice animal house and for the company it was paramount because it's our main activity, we don't have a place without an animal house, we use it daily (interview with Leïla, accredited company director of studies, July 2015).

Thus, this attractive dimension joins the observation of a pragmatic need on the part of the cluster members, who see the use of the platforms primarily as an advantageous service that they can access, which prevails over collaboration opportunities.

4.2.2. *Platform usage: service provision before collaboration*

The platforms offer two different types of usage of their equipment. They offer: (1) service or development activities in support of research laboratories or their spin-offs; (2) they also develop collaboration activities, the results of which are embodied either in coauthored publications or in joint patents:

> The framework contracts are designed to ensure that the benefits from the results are shared. In concrete terms, for the researcher, this means that he or she will be a coauthor of a publication because there is a minimum investment in time, as well as in the use of equipment, so, yes, this can lead to academics being published. Then, in theory, it can lead to

joint patents (interview with Emmanuel, laboratory teacher-researcher and platform manager, February 2017).

These uses echo the typology drawn up by Hatchuel, Le Masson and Nakhla, who distinguish between two main devices (Hatchuel *et al.* 2004). The first is the shared analytical device (*Dispositif d'Analyse Partagé*, DAP), which they describe as a set of rather routine activities: the protocols are already finalized and well known and have a scientific research operation function. They are part of a service provision logic. The second is the shared experimentation device (*Dispositif d'Expérimentation Partagé*, DEP) which, as its name indicates, is centered around experimental and collaborative work. In this case, the research protocol needs to be designed with a high degree of cognitive and geographic proximity to meet the evolving needs of the research object. These two models (service and collaboration) reflect two poles of activity, one oriented towards the commercialization of the knowledge production system, the other towards coproduction and scientific cooperation. The DAP system would then consist of an instrumental link in the subcontracting of R&D. Producing little innovation by itself, a DAP platform is thus distinguished more by its quality service activity with a global network of clients. The main risk it faces is the rapid standardization of the platform in the face of increased competition from more innovative platforms, whether public or private (Aggeri *et al.* 2007a, p. 31).

In 2017, on the biocluster, there are four collaborative platforms, whose utilization is conditional on the development of a joint scientific program, and six service platforms that do not require scientific collaboration. In addition, there are eight mixed platforms that offer both service and collaboration. As we have seen, the technical services of the incubator differ from the rest of the platforms in that it does not offer the provision of specific skills, but rather provides the incubator tenants with access to all of the shared facilities (central laundry, freezer room, sample preparation room, etc.). The dominant use of the platforms seems to be equipment usage in a service rather than cooperative sense. Of the 102 people who took the questionnaire, only 22% answered positively to the question “Have you ever used one or more of the shared platforms?” Half of these were PhDs and the other half engineering postgraduates, PhD students and R&D management positions. 55% answered “no” and 23% said they did not know about the existence of platforms.

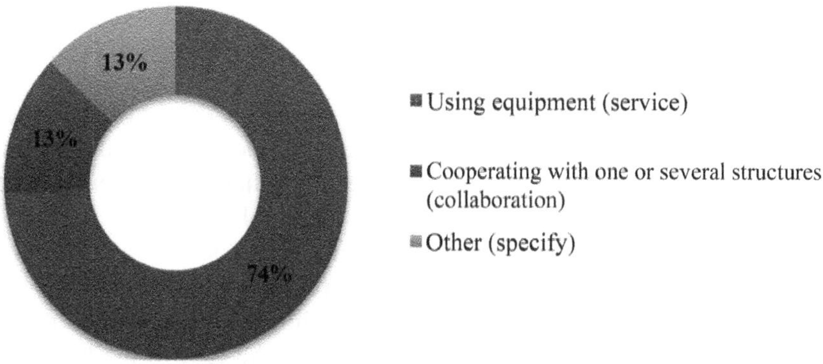

Figure 4.2. *Type of platform usage by users (source: graph constructed from questionnaire data sent to 102 people working in the cluster between February and April 2016). For a color version of this figure, see www.iste.co.uk/vallier/clusters.zip*

In 74% of cases, the panel's users only used the equipment for service provision, in 13% of cases to cooperate with one or more organizations and the remaining 13% who checked the "Other" box all specified that they used it because the equipment was located in their laboratory. From the management perspective, 44% said that their organization made no use of the platforms and 3% did not know what they were. All of the managers who said their facility did not use platforms said that they had no use for any of the equipment offered. Of the 53% who said they did use them, 11 were company directors and 6 were laboratory directors. Fourteen said they used the equipment as a service, only two used it in collaboration and one because it was in their laboratory. In the end, both the quantitative and qualitative surveys showed that the activity is more about service than collaboration. The following statements clearly illustrate this logic:

> I met with the platforms manager at Genopole, at one point, because I was responsible for an L3 laboratory and we wanted to potentially open it up to the outside world if people needed an activity [...] and we have people from many companies who come here, so it's working quite well [...]. But no, it doesn't allow us to meet each other, but it provides a service, because we don't even see the people, they come at their time slot, analyze their data and leave, because it saves them from developing a platform that costs an arm and a leg. We have a

> microscope in imaging that costs 200,000 or 300,000 euro (interview with Fabrice, accredited R&D laboratory researcher).

From this point of view, Genopole appears to be a special case in terms of the way in which the sociology of science tends to identify shared platforms: as a specific trading zone, that is, a zone of exchange between different cultural and disciplinary groups (Galison 1997; Shinn 2007). In France, Vincent Simoulin, in particular, shows the emergence of an interstitial community linked to the diversity of actors around a scientific facility – the first third-generation synchrotron – while highlighting the specificities of this heterogeneous collective (Simoulin 2012). In the case of the biocluster, on the other hand, work on the platform is more compartmentalized than interstitial, as the interview extract above shows. Occasional access to expensive, high-quality equipment is preferred, leading to sessions of usage that do not then allow for sufficient decompartmentalization between research teams or companies in the cluster.

4.2.3. *Equipment demonstrations: connecting or making visible?*

Since 2013, the biocluster has been organizing *Journée des plateformes* (Platforms Day), better known by the acronym "JPF", in the meeting rooms and corridors of Genopole. These days are presented as an opportunity for biocluster members to discover the diversity of the offer and to exchange around needs, partnerships, collaborations, services, etc. These events are part of the action plan implemented by the Animation Manager and the Genopole Recherche Platforms Manager. The aim is to publicize the platform system to accredited members and to extend it to companies and laboratories in the Paris region, which are potential clients. The stated objective is to create a dialogue between the public and private sectors on different topics in order to decompartmentalize research. Each day's program generally consists of a presentation of certain platforms, with priority given to new ones where appropriate. Round tables are also organized on issues specific to the platforms: data security, contractualization aspects, etc. Of the laboratory or company managers who took part in the survey, 56% said they had already participated in the Platforms day, but only 9% had developed cooperative ventures from them. This data is consistent with the testimonies of several platform managers who explained that they had not developed any collaboration during these events:

> We haven't, no, because the themes are really different. Well yes, I'm saying that, but, indeed, I did meet some people who I had never spoken to. However, in terms of the activity of the platform, the JPF does not bring us much, but it is very good that it is there on the site and it is a time when we can see each other. But if we look at the indicators, or in a quantitative way, it has very little impact on our activity (interview with Gwenaël, laboratory research engineer, February 2017).

The above extract reflects a divergence of themes among the participants that is too broad for generating common interests. At best, opportunities may present themselves, but not materialize. Indeed, there are many accounts of "near-relationship" cases, as the following extract indicates:

> So we participated in all these days and, once, we almost created links. Moreover, it would have been great, with L'Oréal, so someone who came to participate in the days but outside the cluster [...]. But we didn't hear from them, it's true that the person we met was extremely interested, but I guess she didn't manage to pass on the message to her managers afterwards. So it's been all downhill from there and, after that, I haven't wanted to chase after industrialists, if they are interested, we talk, but that's it... (interview with Edith, laboratory CNRS research director , February 2017).

This potential collaboration with L'Oréal, which could not be realized, led to a closer analysis of the list of 156 participants at the 2016 Platform Day, in order to identify their institutional affiliation. It emerged that 50% of the participants were industrialists or academics from outside the cluster. Of the other half, 35% worked in the laboratories, 8% in companies on the site and 6% in Genopole.

This attraction of actors from outside the cluster for the Platforms day can be explained by the presentation of equipment by recognized equipment manufacturers. In fact, they were the sponsors of the day; in exchange for their financial contribution, they were able to come and present their instruments, or even give demonstrations during the Platforms day or during dedicated "equipment demo" sessions. The aim of these events was to promote the use of existing platforms through specific workshops or equipment presentations, as well as to bring in instrument demonstrators for future investments. These "demos" are a term commonly used by a multiplicity of actors, including researchers, engineers and computer consultants, to refer to a presentation of a technological device in action, such as software or a robot (Rosental 2017, p. 75). These demos are widely used in Silicon Valley, not only to promote science and technology, but also because they serve to connect multiple actors whose likelihood of coming into contact would otherwise remain low.

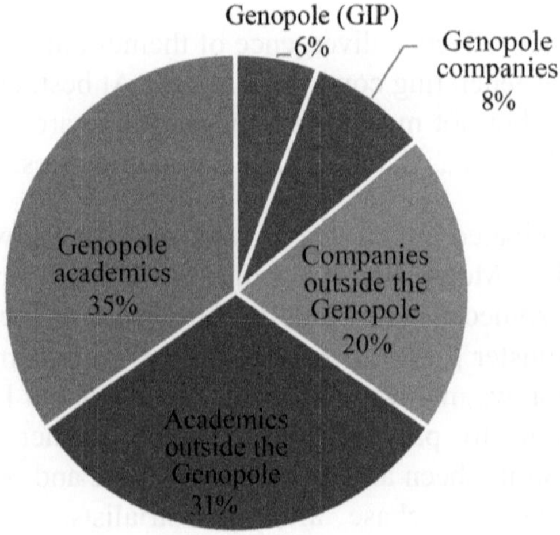

Figure 4.3. *Institutional affiliation of the 2016 Platform Day participants (source: graph constructed from the attendance sheet of the 156 participants attendance sheet of the 156 participants of Platform Day on June 23, 2016). For a color version of this figure, see www.iste.co.uk/vallier/clusters.zip*

On this point, however, as with shared platforms (see section 4.2.2), the biocluster tends to stand out. Rosental, in his survey of the California cluster (Rosental 2017), sees demos as sales tools, forms of self-promotion and interaction. He explains that the "demo" becomes a kind of calling card after a first contact and that thus the spectators are, with time, likely to become project partners, if not users, sponsors or customers of the prototypes or future products presented (*ibid.*, p. 76). From then on, the demos turn out to be operators of networking that replace other forms of social relations between scientists, such as meetings, informal encounters, café visits, etc. (Ravelli 2007). They also empower the demonstrators to develop an experimental approach by observing the reactions of the audience around the objective that the equipment represents (what may or may not be relevant, what needs to be improved, etc.), which makes it possible to "define a common identity and the nature of the relationships between the protagonists" (Rosental 2007, p. 152). This networking function of the demos, as highlighted in the literature, contradicts the results of the questionnaire, where only 14% of individuals interviewed had ever previously participated in one of the demos on the site, none of which had otherwise resulted in collaboration. The following interview extract makes this clear:

> When they bring in equipment manufacturers, people who present new technologies, not many people come. And yes, you can come and see this machine, but in any case we are in companies that cannot afford to buy it and we know that Genopole will not pay for it. So, yes, there have been some, but the machines financed by Genopole are supposed to be pooled, so that they can be used by the companies that have requested them. They say you can come and use them, but in the end you have to come at certain times, within a strict timetable [...]. If you look at the machines bought by Genopole, it's the big companies, they're on their side, and that, here, is a shared feeling and, basically, a feeling of injustice. If we're at X, yes, we have; if we're at the incubator, not so much (interview with Françoise, research engineer in a company located at the incubator, July 2015).

Here, instead of encouraging exchanges, the equipment creates a kind of rivalry between those who can see their structure benefitting from the installation of a shared platform and those who cannot. This feeling of injustice is reinforced by the essential function of this equipment for companies and laboratories. Indeed, the rivalry that these demos bring to light shows the strong presence of fixed capital (instrumentation, platform) in science, without which science cannot exist. Therefore, the demos highlight the visibility and unequal access of the structuring dimension of this fixed capital.

While demos and platforms appear in the literature as potential vectors of cooperation, it is difficult to establish relationships here. The platforms seem rather to be reflective of a policy of making equipment available (in order to ensure the proper development of the site's laboratories and companies) and a perspective of attractiveness. In addition to the support offered, accreditation can be seen as a means of accessing state-of-the-art infrastructure for laboratories and companies.

In this context, networking seems to take a back seat, as the accredited members focus on service use in order to develop their respective activities and not their partnerships. In the context of scientific experimentation that the platforms represent, exchange appears limited. On the other hand, the analysis of knowledge transfer practices also reveals the role of informal exchange spaces (Lamy and Le Roux 2017, p. 109). These spaces are also favored by the biocluster and are part of its palette of tools for establishing relationships, which are conceived in terms of conviviality.

4.3. The institutionalization of conviviality: "*la vie de site*"

The idea of a cafeteria effect was born at the same time as the first technopoles. It is based on the principle that it is during an informal discussion around the coffee machine or in a convivial setting that links are forged and lead to professional collaborations.

4.3.1. *Bringing together and involving employees from different backgrounds*

Since 2010, the cluster under study has been strongly implementing a final dimension for better cooperation, which involves a social and convivial policy around sports and cultural activities. The principle behind this policy is the idea that informal and playful networking can lead to professional relationships. This policy is known as "site life". It refers to cultural and sports activities, as well as infrastructure such as the creation of day care centers or cafeterias. In 2013, the biocluster made this a key focus of its development:

> Administration and site life have been identified as major aspects of the "Genopole 2020-2025" strategic plan. The main objective is to make the Genopole biopark a biocluster defined by the values *Work, Live, Interact and Play*: *Work*: a place where it is good to work; *Live*: a pleasant working environment; *Interact, Play*: providing employees with leisure, cultural and sports activities, etc. (2014 biocluster activity program, section IV. Site Life).

In 2012, Genopole wanted to develop a coherent and integrated strategy for the various user-friendly initiatives developed on the site. To this end, it subsidized a successful local initiative that has accompanied Genopole since its creation. This is a newspaper called *Forum*, created by an employee of the Genomics Institute, which mainly reports on social events such as farewell parties, births and birthdays, as well as on laboratory life (discoveries, technologies, arrivals of researchers, etc.). Created in 1999, this journal was first written by this one employee, then gradually gathered other contributors. For nearly 13 years, this journal was published "by hand" (printed and bound with staples). In 2012, the biocluster committed to financing a more formal edition and recruited its editor-in-chief, provided by his employer.

Forum fulfills the functions of a company newspaper concerning the "family spirit" (Saint-Georges 1993, p. 10) that it disseminates to its readers, using the terms "we" or the "Genopolitans". Above all, it disseminates news from the various entities of the cluster under the same banner, whether they are of a convivial or professional nature. This aggregation within the *Forum* newspaper is indicative of the general strategy of "Site life". Indeed, the editor-in-chief of the newspaper is also the "Site life" project manager within Genopole and, when taking up his post, his mission was to deploy an overall approach based on already existing initiatives. There are several formal and informal association initiatives in the biocluster. These include the Genopole workers cultural association (*Association Culturelle des Travailleurs de la Genopole*, ACTG), which has been operating within the Genomics Institute since 2003 and offers activities as diverse as language courses, guitar lessons, dancing, etc. We can also mention the *Association Le Téléthon des Chercheurs*, created in 2006, which essentially brings together people from the AFM-Généthon-I-Stem sphere, and which organizes various activities around the collection of donations for the Telethon and runs the Évry Telethon Village. It should be noted, once again, that these initiatives depend on the two structuring actors of the site: the AFM and the Genomics Institute. Alongside them, Genopole has also set up a number of initiatives, notably the "Gen'envie" platform[14], which organizes a series of events: an annual picnic in the communal gardens, popularization courses, etc.

---

14 Internet portal for employees (1,400 people) of accredited organizations who pay a fee for access, which allows them to benefit from special offers on tickets, practical information (classified ads, carpooling, accommodation, etc.) and newsletters (at least once a month).

Faced with this diversified but fragmented offer, the need for overall coherence on the campus as well as employee involvement became decisive. This convivial dimension was piloted by a project team (composed of Genopole's structure employees) which proposed an action plan and objectives. The aim was to create an association of employees who would take charge of cultural and sports activities, by getting the employees involved so that they would take ownership of the project. The action plan was inspired by a series of methods, themselves inspired by participative management. Although the literature observes the "lack of clarity in the definition" of this management (Borzèix and Linhart 1988, p. 38), the main device remains the work group in which employees are mobilized around a particular issue or problem. First of all, the project team, having committed itself to providing a multi-purpose room, set about gathering the expectations of the employees concerning the activities they wanted. A questionnaire, with the unifying slogan "Deciding together changes everything", was distributed to all the member establishments of the site. More than 500 people responded to this questionnaire, which ended with an invitation to those interested to take part in workshops, with the title "General conditions of site life", and to leave their contact details. The results of this consultation were then used as a basis for these workshops, which took place a few weeks later. The ideas developed during the workshops were presented in a document, but were not really implemented. Effectively, the working groups were composed of biocluster employees, but with different backgrounds; thus, their participation in an artificially formed group, outside of the work group to which they belonged, strongly constrained their involvement. These resistances can be found in Borzèix and Linhart's definition of participative management:

> The new participatory formulas depart from this natural and private universe. They are organized, initiated and monitored from the outside by the management, on its initiative and under its control. They are artificial (because of the places where they are held, in meeting rooms, outside the workshop); they are foreign to the customs of productive work […]; they are specific or even exceptional […]; finally, and above all, they do not follow […] either the contours or the functions of natural work communities […]. Extracted from their group membership, the individuals who participate in them are confronted with logics coming from elsewhere, with new

> interlocutors: workshops or neighboring offices, functional or administrative services [...] they are invited to abandon their original solidarities and their partisan alliances to leave their confinement and their particularism. They are encouraged to become aware of their belonging to a whole (*ibid.*, pp. 49–50).

This particularism, which was mentioned above, is all the more reinforced in the case of these "General conditions", insofar as the participants do not have the same employers. Thus, in addition to this extirpation of the original belonging, it is very difficult individually to be committed in the workplace to a project that is not that of one's employer, all the more so when this project concerns non-professional activities.

4.3.2. *Building emotional and community connections through volunteering and sport*

The actual involvement of employees in the implementation of the conviviality policy was quickly abandoned in favor of direct responsibility by the project team. On the other hand, the participation of employees was decisive in encouraging them to "become aware of their belonging to a group" (Borzèix and Linhart 1988, p. 50). They were therefore invited to take part in symbolic actions that allowed the development of a common identity for this heterogeneous group. Indeed, following the working groups, a new questionnaire consulted employees on the name they wished to give to the future multipurpose hall. Proposals were suggested electronically and then a vote was held to elect the winning proposal. At first, this type of device emphasized a predominant affective dimension in group relations (Sainsaulieu 1985, 1977 and 1988). Indeed, a recreational moment like this temporarily minimizes the importance of professional group memberships. Then, the principle of the vote granted a democratic spirit to the construction of the project. Reciprocally, this democratic gesture also participated in creating an affective link between employees, who decided on the name, even if not unanimously, and the multipurpose room project. The name finally chosen, *L'Escale*, reinforced the recreational character of the place at the same time as it associated an idea of space outside professional time. This construction of identity was a feature of managerial practices known as "governance by culture" (Crozier 1991, p. 51), or "cultural management", which consists of conceiving a corporate culture composed of values, norms, rituals, myths, etc. The constitution of a community requires a sense of belonging, in which the individual's institutional support is sought:

> In a community issue, consent is no longer only contractual, but also, and above all, affective. Moreover, the feeling of belonging must be so strong that the link that unites the employee to his company must be more an affective and moral link than a contractual one (Barbusse 2008, p. 409).

Thus, the feeling of belonging to a collective – this is the word used by the project team in its communication: "a collective of biocluster employees", initiated during the working groups and then confirmed with the choice of the room's name, is accentuated throughout the project. Indeed, meetings are organized quite frequently, about every three months, before and since the opening of the room, in order to keep the employees involved informed about the progress of work and the organization of activities. Following a call for applications, some employees from the site volunteered to run a workshop at L'Escale. When the hall opened, "open days" were organized with booths run by the volunteer organizers: knitting workshops, fitness, dance and relaxation classes, groups of soccer players and an athletics club. Registration for sports activities was the most successful, but posed problems in terms of insurance, which prompted the collective to become an association under 1901 legislation. Thus, the *Escale des Genopolitains* association was created at the beginning of 2016, one year after the start of activities in the hall. A membership system was set up and very quickly the number of members passed 250. In addition to the workshops, festive and convivial gatherings (afterworks, picnics, etc.) and events (Easter chocolate sales, clothing exchanges, Christmas markets, etc.) have been added. Overall, the desire to involve employees outside of their professional duties is mainly through sports mobilization, since "when we know the unifying and identifying capacity of sports, we understand why companies that want to unite their employees around common values turn to sports" (*ibid.*, p. 409). This ambition is part of a double movement. On the one hand, personalities of the wellness movement are inspiring the emergence of employees eager to start their day with a jog or a fitness session (Cederström and Spicer 2016). On the other hand, the directives in vogue around well-being at work prescribe the installation of table soccer, rest rooms or the organization of afterworks, etc.

4.3.3. *L'Escale, a space of sociability revealing professional hierarchies*

Creating the conditions for conviviality is an old idea, the beginnings of which can be seen in the discourse surrounding the emblematic technopole, represented by Sophia Antipolis in the innovation imaginary in France. Pierre Laffite, its director, explained the need for the Sophia Antipolis Foundation to maintain conviviality on the site in parallel with more professional events:

> We also need more convivial and confidential meeting places so that scientists and businessmen feel they have a home of their own where they can meet for a debate, dinner or show and do business. It is necessary that the baker can meet the financier and the chemist can meet the engineer, that is cross-fertilization [...]. Other meetings take place during lunches, exhibitions or concerts. It is interesting that a music lover from Digital Equipment can meet a music lover from the Ecole des Mines at a Mozart quintet concert, because they can then develop other relationships ("Les leçons de Monsieur Technopole", Meridian interview, 1988, p. 37).

In the same sense, L'Escale is envisioned as a place where individuals can meet by chance, during a sports or cultural activity. The questionnaire survey reveals that the hall was frequented at least once, in 2016, by 45% of the 102 individuals interviewed, making it the most used sociability space at the time of the survey. The literature on this type of space emphasizes its importance as an institution of socialization. The expression is used to describe the places where the French community in Silicon Valley socializes, such as the French school in Palo Alto, the film club, French restaurants, July 14th celebrations or the Beaujolais Nouveau Day (Dibiaggio and Ferrary 2003). From this point of view, this community developed behaviors specific to any expatriate ethnicity, by reproducing the characteristics of French culture and practicing its mother tongue (Portes 1995). Although the ethnic character is not appropriate in the analysis of L'Escale, it participates, like the French community, in the construction of non-economic social ties through communal activities. Nevertheless, in the case of L'Escale, it is difficult to speak of socialization, because it is not a matter of individuals internalizing or sharing a value system. It is more appropriate to refer to it in terms of a space for sociability. Indeed, its success has been explained, on several occasions, by the activities proposed and by the fact that it represents one of the rare places of sociability:

> L'Escale, clearly L'Escale. So it's all new, I tried Shiatsu, stretching and fitness [...]. There are people I know from here (the nursery) and we meet at L'Escale and we talk more easily. Afterwards, I'm a fairly sociable person, so I don't mind sitting at the table and talking. I know some people from the company who will go to L'Escale, but who will stay in their own little group and, outside of L'Escale, there's not much socialization (interview with Leïla, accredited company director of studies, July 2015).

The point raised in this extract, the formation of groups on a professional basis during convivial events, came up several times during the survey. Indeed, although they are away from their workplace, whether at L'Escale or at picnics, employees tend to recreate groups based on their professional affiliation:

> I only came to one presentation where investors were there, my manager went more often. Me, I went more to the lunchtime events, there was a picnic, there were a few tips... The second time, I went with the idea of getting to know people and, again, it's not easy because people know each other and stay in their own companies (interview with Rizlaine, accredited company biomechanics engineer, December 2014).

This biomechanics engineer, the only employee in her company besides her manager, was keen to meet people over lunch, but found herself confronted by groups already formed within their own companies. Moreover, she mentions a dichotomy in social events, some of which were more reserved for her manager. More generally, this dichotomy illustrates the fact that some events involving professional time are reserved for those in positions of responsibility, while others, more convivial in nature, are aimed at the "general public" during lunch hours. Thus, certain conferences or meetings related to scientific and professional activities would be difficult to access for engineers or technicians who do not seem to be inclined to represent their company. However, L'Escale and other related events would bring everyone together. In reality, however, the survey shows that L'Escale is a place that is little used by managers and people in positions of responsibility.

Consequently, the low investment of management in this "non-market" space does not participate in the mode of governance implicit in the coordination of actors that has been pointed out in the literature (Saxenian 1994). The following extract testifies to this:

> I think it can lead to something if the bosses are in that informal relationship; if a technician becomes buddies with another technician it won't lead to anything (interview with Leïla, accredited company director of studies, July 2015).

For this director of studies, the informal exchanges likely to become professional collaborations take place between managers. Thus, coordination resulting from the use of conviviality places cannot be expected if they are absent. As a result, despite the constantly increasing number of registrations, L'Escale appears to be more of a service available to the biocluster's employees than a true arena for the construction of non-economic social links that could be converted into relationships with an economic purpose. In this context, it appears difficult to compare the economic potential of L'Escale to that of the French community analyzed by Ferrary, who assures that, if a French person in Silicon Valley were looking to create a company, he/she would spontaneously turn to his/her ethnic community for advice and information during social gatherings.

As we end this chapter, we can see that cluster workers seem to encounter difficulties in terms of availability in order to participate in the various networking events, and, even when they do have the opportunity, the exchanges are rarely converted into professional collaborations.

5

Scientific Competition and Economic Competition: Social Fields Spanned by Internal Struggles

As we have just seen in the previous chapter, most clusters, most notably the one studied in this survey, have set up multiple networking mechanisms. In spite of all these initiatives, it is difficult for cooperation to emerge. In fact, they have to emerge in a context of scientific competition between laboratories and economic competition between businesses. This competitive context is reinforced by the constant search for funding for projects and by the growing insecurity of researchers and entrepreneurs alike. Moreover, spontaneous collaborations between science and industry are developing slowly despite geographical proximity, mainly because of differences in work organizations and between disciplines; and this, despite unifying names such as "biotechnology".

5.1. Asynchronous organizations and work rhythms

The diversity of the organizations gathered in the cluster is a source of divergences and stumbling blocks, particularly through the temporalities specific to each type of structure and their sometimes very distinct disciplinary groundings.

5.1.1. *Dissonant work schedules between companies and laboratories*

In all of the organizations observed during the survey, time lags, particularly between laboratories and companies, were revealed. In this respect, one company director confided the difference in approach to a CIFRE thesis between her company (which employed the PhD student) and the partner laboratory located in the cluster (which supervised the scientific process):

> As a result, this is really felt in the contractual exchange that we have with this CIFRE thesis. On their side it's cool. We need to make progress within six months in order to market something. It's really two different ways of looking at the thesis... (Interview with Sylvie, director of a company accredited by the biocluster, June 2015).

This interview extract highlights a gap and scheduling problems that may emerge from this shared postgraduate research, where the PhD student is at the interface between science and industry (Levy 2005). These differences in temporality are generally found, between science and industry, at the time of cementing relationships:

> Probably because what we do is still very academic. There is a gap to be bridged between the results we produce and those required by industry [...]. We are not at all in an industrial process. We test enzymes and substrates. If they are varied, we find activities that are more or less significant, or may or may not be marginal. Afterwards, what are industrialists going to do with these enzymes? [...] There is a whole development process that follows, that we cannot assume, because it is not our job [...]. Manufacturers want an immediate response to their needs. That we are not able to provide (Interview with Patrick, research institute director, September 2015).

These words echo the work that shows that universities and companies are founded on different systems of action and norms, with open science evaluated by peers, on the one hand, and proprietary science evaluated by the market that seeks to profit from scientific results, on the other hand (Dasgupta and David 1994). Although theoretical models have long been used to describe coordination mechanisms between scientific and industrial communities as “boundary objects” (Star and Griesemer 1989) or “transaction fields” (Galison 1997), the Genopole survey and the network analysis carried out reveal phenomena of incommunicability. These are directly linked to different visions and ways of doing things. Science has a primarily operational value for industry, whereas in academic research, even though the market is increasingly present, researchers maintain an ethos marked by the universal and general interest nature of knowledge (Merton 1996). Moreover, the university pursues three missions (education, research and valorization), but the evaluation of its workers focuses on research that has resulted in publication (Gastaldi and Lanciano-Morandat 2017).

On the industrial side, the speed with which they must renew themselves in a globalized economy leads them to abandon certain scientific applications in the search for immediate profitability (Uzunidis 2003, p. 69). The example of shared platforms (see Chapter 4) shows that, although they are shared workspaces, they merely accommodate the different professional logics of the organizations accredited by the cluster. In particular, this is a domestic use for laboratories, which allows them to have access to expensive equipment. This use is also valid for the companies in the cluster, but coupled with an opportunistic use of equipment for commercialization purposes (Aggeri *et al.* 2007a, p. 121). Moreover, the timeframes also seem to differ between firms, depending on whether or not they are in the development phase. In this regard, let us listen to this company director who suggests a categorization of structures and associated logics:

> We all have our own agendas, our own way of doing things, so we don’t work together [...]. Perhaps, also, because, as they are start-ups, they are often chasing the same markets. There is a kind of rivalry between them, competition for investors [...]. We’re a little bit different... You’d have to segment the Genopole into pure start-ups, pure university labs, commercial enterprises, etc. [...]. I think everyone works very differently (Interview with Philippe, director of a company accredited by the biocluster, December 2014).

As a result, the institutional logic of the cluster, which seeks to bring the different logics together, comes up against this segmentation. The cluster’s intermediation structure does not escape its own administrative timeframes either, which appear to be unsuitable, especially in the eyes of companies:

> Genopole is structured, but structuring means administrative slowness. Clearly, the operation we were able to set up in 2002 was completed very quickly, in four months. If the same operation were to be launched today, it would *a priori* take at least a year. So Genopole's responsiveness has slowed considerably [...]. And the reality is that a biotech company has to make decisions extremely quickly. Genopole's timelines are not compatible with what Genopole can offer [...]. Everything has to go through tenders, etc. (Interview with Alexandre, accredited company director, September 2015).

Beyond the bureaucratic considerations linked to the organizational structure of the cluster, this company director also stresses that Genopole's objectives may not match those of the accredited members, as in the case of the biocluster's strategic orientation committee (*Comité d'Orientation Stratégique du Biocluster*, COSEB) which brings together company and laboratory directors:

> Originally, COSEB was set up without any specifications, but little by little Genopole directed it towards defining the Genopole 2025 strategy. However, the people on COSEB did not have the skills to define what biotechnology might look like in 2025, given that they were thinking in terms of two to three years, not ten to fifteen. So there was a completely incompatible delta between what Genopole wanted and the people who were elected to the COSEB (Interview with Alexandre, accredited company director, September 2015).

In this extract, we note the distinction between the temporal perspective, in which the public intermediation structure is positioned, and that of the COSEB representatives where, as a reminder, 8 out of the 17 elected members are company managers.

Finally, we note another temporality within a structuring actor of the cluster: the French Muscular Dystrophy Association (AFM). The various interviews and observations conducted within the association show a particular attachment of employees to their institution. Many of them consider that they combine their professional activity with their activism in the fight against Muscular Dystrophy. In this respect, the Telethon is an event that punctuates the life and temporality of the association every year.

From September to January, employees are highly mobilized and have little time for outside exchanges. Let us listen to a communication officer who expresses the time and moral investment:

> When you come to the AFM now (in July) and when you come to the AFM in October, there are two different worlds, there's a real pressure behind it. I arrived in August, so I had three weeks of integration and from September onwards it was crazy! It's too much pressure, you know that there are children waiting for you to do better, we're not alone, I mean, you know that behind it there are lives […]. And then you have key events during the year that remind you why you are there. There's the Telethon, you're ready, you're all over the place, you're stressed, you don't really see what's going on, you don't really take the time to talk to people. Whereas on Family Day in June, you have all the families around you, you can talk to them and then you understand why you are there; I think that's also what keeps you going (Interview with Jeanne, communications officer at AFM, July 2015).

In this interview extract, we note that in addition to the Telethon, which takes up the teams' time over several months, the AFM regularly holds other federating events in the fight against muscular dystrophy, such as the Family Day. This event is used as a reminder of the purpose of the moral commitment of employees to the families of sick children.

In the end, we observe a desynchronization of temporalities that is particularly embodied in the distinction between the scientific time of laboratories and the commercial time to which companies are subject. In addition, alongside these two structuring temporalities, we also observe associative time, through the activist nature of the AFM, which imposes a particular annual rhythm with the collection of donations, and the administrative time of the intermediation structure centered on the issues of convergence between differentiated actors. It is also possible to add the medical time of the nearby Southern Paris Region Hospital (*Centre Hospitalier Sud Francilien*), where the priority remains care. Indeed, the few interviews conducted at the hospital show that the temporality of the institution is dedicated to the care of patients and dissociated from entrepreneurial and scientific logics.

5.1.2. *Belonging to the large biotechnology family or disciplinary demarcation logics?*

In addition to this temporal dissonance between distinct social fields, all of the structures seem to be confronted, whether between laboratories and companies, within companies or within laboratories, by a distinction between fields of activity, which would partly explain the lack of cooperation between structures. Indeed, the lack of common interest is materialized, on several occasions, by the fact that the disciplines are too distinct:

> I think the themes are so distinct that the interactions are quite difficult... Let's say we have an approach... We work on plants and we are the only ones. After that, what can bring us closer together is the high-throughput sequencing aspect, from a technological point of view, where there may be certain similarities. The others are very health-oriented and not so into plants (Interview with Maximilien, accredited laboratory research engineer, November 2014).

This distinction between disciplines is particularly noticeable in the accredited organizations that do not work in health-related biotechnologies, which are the cluster's original domains. As a result, individuals who work in different fields do not feel included in the seminars and conferences organized:

> We're focused on agri, whereas Genopole is, broadly speaking, focused on human health. As a result, most of the seminars focus on molecular biology, quite heavy stuff, health approaches, etc. (Interview with Nora, marketing manager at a biocluster-accredited company, November 2014).

This company, which develops technologies based on plant genetic manipulation, does not see its place in the overly generalized field of biotech. Without it being possible to establish a definite cause and effect relationship, this company left the biocluster 2 years after the interview with this marketing manager. The distinction between disciplines is not only felt in companies and laboratories working in the plant sector. It can also be felt in organizations in the health field:

> I was wondering about moving to Cochin. They have a business center and, when you see the list of companies, these are people you can talk to because it's clearly biomedical [...]. Genopole is about genes and cells and, in fact, it's a bit of everything. There are people who make viruses, therapeutics, molecular analysis kits, etc., so we don't interact with them much (Interview with Sylvie, director of a company accredited by the biocluster, June 2015).

The "little bit of everything" in this director's words is symptomatic of public research policies that promote interdisciplinarity. Indeed, LABEX (Laboratories of Excellence), IDEX (Excellence Initiatives) and even the recent I-SITE (Initiatives for Science, Innovation, Territories and Economy) are emblematic of these interdisciplinary policies to bring science and industry closer together. These public innovation policies are reconfigured around "major programs articulated around keywords whose definition is elusive: 'nanotechnology', 'biotechnology', or 'neuroscience' and 'genomics'" (Fallon and Thoreau 2016, p. 17). These very broad headings cover different practices and disciplines that share a relatively low common denominator:

> The principle of Genopole is to say, "We're going to bring in good scientists, we're going to put them together and we think they will gel". But that's not true, that is not how it works, if you just take the example of bioinformatics, there are procedures, methods, tools that are specific to statisticians. They don't have the same brains as geneticists! So, can you imagine when you have to explain what you do to a guy who only talks about business? (Interview with Gérard, accredited laboratory teacher-researcher, May 2014).

In the case of biotechnology and genomics, the common denominator is the manipulation of life. But, as we saw in the previous extract, misunderstandings can emerge between biostatisticians and geneticists, even though they all work on living organisms. Some even speak of a "gulf" between the different activities of the site, which would greatly limit the mutual interest in creating collaborations. From this point of view, it appears that the objective of restructuring public research as a whole, organized around major application areas (health, energy, etc.) to the detriment of disciplinary research (Hubert and Louvel 2012), is meeting with resistance.

These disciplinary qualifications, such as biotechnology, or identity qualifications, such as the creative class, therefore mask a diversity of fields of activity and different individual and collective practices. The observation developed here is largely in line with the position of Jean-Pierre Durand, who sees in Richard Florida's creative class, not a "class in itself" with similar conditions and lifestyle, but an amalgam of intellectual occupations (Durand 2016). Moreover, the professionals of the creative class would, according to Durand, still not be aware of a common belonging and would therefore not constitute a "class for its own sake". This idea tends to be corroborated by the survey conducted at Genopole. Although the majority of the cluster's workers are highly qualified, their positions in the social space, as well as the social fields to which they belong, differ greatly. As Jean-Pierre Durand rightly says, Florida adopts a substantialist definition of social classes, based on the content of their members' activities. The following two points are intended to shine a less homogeneous light on the different workers studied, who will therefore be analyzed according to the field to which they belong and the position they occupy within that field.

5.2. A scientific field built from struggle and precarity

The scientific field has undergone a metamorphosis over the last 30 years, with a decrease in tenure, project-based funding, industrial collaboration and so on, the consequences of which are felt in the laboratories of the cluster.

5.2.1. *A workforce that is becoming precarious*

Anti-differentiationist theses have developed in opposition to that of Robert K. Merton (Lamy and Shinn 2006, p. 24). In Merton's conception, science is seen as a social structure in its own right, made up of norms and modes of regulation, thanks to the concept of ethos, which makes science an institution distinct from the rest of society (Merton 1996, p. 223). He asserts that scientific ethos accounts for a scientific community animated by the development of science. Gradually, work that took science as its objective called this approach into question in favor of developing the idea of science as a pre-capitalist market activity (Hagstrom 1965). This individualistic view of scientists emphasizes the search for recognition by peers. Hagstrom sees, in the commitment of researchers (contribution to a journal or to a communication), a gift that calls for a counter-gift (recognition). Although far from individualist paradigms, Bourdieu also breaks with the idea of a scientific community driven solely by a common objective of scientific progress. Through his notion of social field, he views the scientific community as a space of struggles between researchers. He sees science as a social field like any other, with power relations, strategies, interests and so on. Researchers are therefore in competition to obtain, not just recognition, but the monopoly of scientific authority. The field of science is therefore understood as follows:

> A system of objective relations between acquired positions, the scientific field is an area of competitive struggle, whose specific stake is the monopoly of scientific authority, inseparably defined as technical capacity and social power, or, if one prefers, the monopoly of scientific competence, understood in the sense of the capacity to speak and act legitimately (in other words, authoritatively and with approval) in matters of science, which is socially recognized to a determined agent (Bourdieu 1976, p. 89).

In this vision of the scientific world, researchers produce and interact in order to increase their social and symbolic capital. To scientific recognition and authority, Latour and Woolgar add credibility in terms of the ability of researchers to convert one capital into another, for example, the conversion of scientific facts into grants (Latour and Woolgar 2006). Like capitalist entrepreneurs, the greater the accumulation of a stock of credibility, the greater the researcher's ability to increase his capital.

If we continue the Bourdieusian reasoning in terms of a social field, within these struggles and power relations, we see the appearance of the dominators and the dominated. In his thesis (Heil 2010), Christophe Heil identifies two types of workers in university research laboratories, both of which are affected, in differing degrees, by the growing insecurity of university staff: the "long-term temps" and the "insecure academics". The former are characterized by the accumulation of statuses that usually "should only be temporary leading to permanent positions, they follow one after the other until they finally end up at a professional dead end" (Heil 2010, p. 355). He thus refers to the statuses of fellow, lecturer, training and research assistant (*ATER*), paid unemployment at the end of the ATER contract and a number of postgraduate positions. While it was less present in

France than in the United States, the use of postgraduates is becoming more widespread, particularly in the life sciences (Hubert and Louvel 2012, p. 23). This extension of short-term status and limited remuneration can also be observed in the cluster, mainly through the large number of people with postgraduate degrees. Of the 22 PhD graduates who responded to the questionnaire, 14 completed it with a post-doc. The same is true for the interviews conducted with 12 PhDs, half of whom also had a post-doctorate. This succession has become the norm and freezes, for years, precarious statuses that were imagined as transitory. During an informal exchange, a doctor in biology even mentioned the obligatory nature of this status: "Without a post-doctorate, you can't have a career in research", he said. The tenure track then moves backwards from year to year. For example, at the CNRS, the average recruitment age for researchers has increased from 32.9 years in 2007 to 34.2 years in 2015[1]. Temporary workers represent up to two-thirds of the teaching staff: "Scientific research only stands because it relies on an army of contract workers, temporary workers and out-of-work doctors" (Granger 2015, p. 13). In particular, 2014 saw two trials emblematic of the situation of these insecure researchers. INSERM was condemned by the Nantes Administrative Court for not having renewed the position of an engineer after 11 years of successive fixed-term contracts. The institute was obliged, following the lawsuit, to proceed with a tenure. The same fate befell the Institute of Myology, the same year, which did not wish to grant a tenure to one of its researchers who had accumulated 7 years of temporary contracts.

These phenomena are particularly the result of a new division of labor linked to project-based funding (see section 5.2.2), which implies that permanent staff develop projects and seek funding while temporary staff move the scientific tasks of the laboratory forward. The latter then see their activity remunerated by the research contracts obtained by the former (Hubert and Louvel 2012, p. 23). In this system, if it works at all, the eventual tenure of young PhDs often depends on the arbitration of the incumbents and can thus encourage clientelistic logics, as explained by a bioinformatics PhD student encountered during the survey:

> At first I went into academic research for the sake of science in the noblest sense of the word, if you will, and in fact I discovered a whole world… where it is the race to publish and

1 https://www.cnrs.fr/sites/default/files/download-file/cnrs-en-chiffre-2016.pdf.

> network to secure a position. I was just discussing this with a friend who is not in that field at all and I was saying that it is the "prostitution of research"! (Interview with Anastasia, bioinformatics PhD student in an accredited laboratory, September 2015).

In this competitive context of access to positions, young PhDs are increasingly turning away from a scientific career, given the involvement in the work and the time availability that the path to tenure implies. In Belgium, the experienced or expected difficulty in balancing work and family life would dissuade 50% of female PhDs and 27% of male PhDs from continuing in a scientific career (Meulders *et al.* 2012). Moreover, the recognition and professionalization of the status of worker in laboratories is a particularly long process. Through an ethnographic survey of chemistry and biology PhD students, Louvel shows that they find themselves in formalized hierarchical relationships, obliged to work in the laboratory, at their bench and respecting the institution's schedules (Louvel 2006). In this context, they are seen more as future colleagues than as young researchers in their own right. Indeed, they are considered as students who will sooner or later integrate into the research environment, while paradoxically they assume a large part of the laboratory's work. In fact, PhD students in the life sciences represent 30% of the qualified workforce mobilized in the scientific production of laboratories, whereas only 25% of them are recruited each year after receiving their doctorate (Mangematin 2003).

As a result, capital and habitus are different between permanents and contractuals, which gives rise to a struggle similar to that observed in private companies, where the dominated "peripheral employees" (fixed-term contracts, temporary workers, involuntary part-time work, etc.) try to join the dominant group, the "core employees" (permanent contracts and full-time workers) (Durand 2012). Thus, the group is far from being homogenic, admittedly subject on the whole to budget reduction logics, but whose consequences vary significantly according to status.

5.2.2. *Scientific work destabilized and concealed by competition*

For their part, tenured staff benefit from a strong statutory protection, but are also affected by a certain insecurity, linked to a double movement of managerialization (performance, competition, productivity) and bureaucratization (procedures, budgetary control, etc.) in research and higher education (Gastaldi and Lanciano-Morandat 2017). Since the creation of the French National Research Agency (*Agence Nationale de la Recherche*, ANR) in 2005, they have been subject to the logic of calls for projects. The annual setting of research themes deemed to be priorities has lead to a generalized competition between researchers on the national market for research funding. This requires skills that are not strictly scientific and that not all researchers have mastered. Responses to calls for proposals are no longer considered as a bonus, but as a prerequisite for scientific practice. This system progressively weakens the members of the academic field who are forced to seek external funding in a context of rapid turnover of calls for projects. However, it seems that it is more a question of a transformation in the realization of scientific work than social and economic insecurity of these workers, whom Heil calls "insecure academics". Indeed, their position is not threatened, either in terms of their remuneration or duration, which is not the case for ATER or postgraduate positions. On the contrary, the system of calls for tender or projects does shape the way research is approached. Indeed, this form of project-based work gives way to "managementism" that encroaches on scientific production, as workers are forced to be accountable, through administrative procedures, and leads to short-term applied research (Goussard and Tiffon 2013). Moreover, the project must meet the requirements formulated by the funding agency and gives rise to a series of phenomena that constrain the researcher even more. The deadlines' and conditions' set effectively disrupts the researcher's work rhythm and their determination of scientific questions. Moreover, the demand for results leads them to detail the stages of their work, whereas research is intrinsically linked to uncertainty. Finally, this demand pushes them to investigate fields that are already marked out, which minimizes risk-taking. Some see the disappearance of a system based on recurrent credits as making it even more difficult for researchers to (re)originate new theories (Heil 2010, p. 391).

Project-based funding is effectively replacing the collective budgetary policy of laboratories, that is, the common pot that allows management to decide collegially on the orientation of resources (Hubert and Louvel 2012, p. 19). This changeover then individualizes the funding that will be granted to a researcher or to teams that have mastered the code of the call for projects. It therefore promotes the emergence of management professionals. Researchers are thus individually led to become "competition bureaucrats" (Granger 2015, p. 106) and managers of their "small academic company [...]

with a budget to spend, a team to work with, deadlines to meet, and reports to compile" (Gastaldi and Lanciano-Morandat 2017).

In addition to calls for proposals, these values of competition and rivalry are fostered by the increasingly important role played by performance indicators that disrupt the scientific ethos (Piron 2008). These indicators cover multiple dimensions, such as productivity in terms of publication, measured by journal rankings or by the calculation of bibliometric indicators. This competition is in action in the scientific work practices of the cluster studied. The constitution of a group of PhD students from different laboratories at the time of the survey highlighted the instructions that were given not to divulge any information. Indeed, students receive official instructions from their laboratories, formalized in the signing of confidentiality agreements concerning their results, which stipulate that only published results can be presented to the outside world. In addition, unofficial instructions from their superiors are distilled within the teams. In particular, the students report the strong competition that exists within teams and which leads to the presentation of only positive results that condition the funding of the teams at departmental meetings. The results are therefore concealed, which, at best, leads to waste (duplication of vectors, antibodies, non-sharing of protocols) and, at worst, to cases of sabotage such as cellular contamination. The phenomena of withholding information, between laboratories and also between research teams belonging to the same institution, are thus very present. Some research team leaders give instructions, particularly to young researchers, not to reveal the progress of their work to their colleagues in neighboring teams. This is done through several strategies: coded presentations in front of peers, even the dissemination of false information, or the prohibition to participate in conferences without hierarchical authorization or without having published beforehand:

> If I want to talk about something outside, we have a disclosure form. So we have to say, "I'm going to present this at a meeting, it's already been patented and we're about to publish it', so there it's validated. If it's new and I've never talked about it, it can't come out, but, even if I'm going to make a presentation in front of collaborators, I have to say that I'm going to disclose something, it goes in front of the R&D manager. This is new, in the last year, because they realized that people were going to talk without saying so. So, now, it's

> controlled. Now, when we're going to present to people who may be interested, we make them sign a CDA (Confidential Document Agreement) and the person must not talk for several years about what I told them. We are very careful […], but it has to be clear, there has to be a procedure to follow, which is what we have here, but does everyone have it? In any case, everything has to be formal and clear, even with Genopole companies. There is no question of doing anything that would upset our intellectual property […] and then, as an individual, I could be fired for passing on confidential information […]. We used to say, "we are going to have a meeting with lab X, where everyone is going to present their activities", but we coded everything… It is not possible because I am presenting to people I do not know who are outside the company and they were coding too. Clearly, all presentations were coded and yet we are in the same sphere, but we are not the same company, so we have to be careful (Interview with Fabrice, researcher in an R&D laboratory accredited by the biocluster, January 2015).

Thus, we observe individual and collective strategies for retaining information, between laboratories and between companies, but even between research teams within the same institution, in order to obtain funding. In this context, interindividual relations at the cluster level can be constraining for researchers who compete or for those who take on increased administrative tasks. In addition, this context can exacerbate tensions related to position and status. Insecure staff, for example, may prefer to value their work rather than participate in collective laboratory moments (Gastaldi and Lanciano-Morandat 2017).

At a collective level, laboratories are also subject to various evaluations and indicators (HCERES, global university rankings, cluster accreditation, etc.). This competition promotes the idea that larger research laboratories are preferable (Vinck and Zarama 2007, p. 278). With sufficient critical mass, they would carry more weight with respect to Europe or the Ministry of Research, which fund them. We then observe the merger of laboratories, which modifies and shakes up the working practices of teams that were previously different, as this laboratory director anticipates:

> In fact, because of budgetary questions from the Ministry of Research, the lab is at risk of being grouped with one or maybe

> two other labs. If that happens, it's going to take up a crazy amount of my time and, above all, we won't have any visibility on the reorganization that will follow: will we have the same type of funding? The same work? The same team? We have a lot of questions (Interview with Blandine, accredited laboratory director, October 2015).

We see in the words of this researcher the fear of seeing organizational logics change and the apprehension of the time that she will have to dedicate to this adaptation. Indeed, as Vinck observes in his ethnography of laboratory mergers, the resulting research collective is the result of a difficult construction of new operating rules, where managerial and scientific logics clash (Vinck and Zarama 2007). Moreover, scientific budgetary policies seem to prevent certain laboratories from using instruments or products that are commercialized on the market, due to a lack of resources. The cofounder of one biocluster start-up explained that, despite his partner's previous research activity in one of the largest laboratories on the site, his company was unable to sell diagnostic kits that he considered useful for the laboratory's activity:

> Yet, we solved specific problems at this lab when [his partner] was there. I mean, gee, why don't they buy our stuff? That's so stupid! The reason is that they are so caught up in budget restrictions and lots of complicated things that, in the end, they don't have time... (Interview with Michel, director of a company accredited by the biocluster, December 2014).

While the work of researchers increasingly requires the use of specific equipment or products in the course of their research (Joly 1997), they are also increasingly confronted by budget constraints. In addition, they are increasingly encouraged to work in collaboration with private actors.

5.2.3. *Researchers and industrial collaboration: an unequal commitment*

In addition to bibliometric indicators, positions in world rankings, holding accreditations, research quality and consequently the ability of laboratories to obtain funding also depend on their ability to work with other disciplines and to collaborate with industry. The intensification of publication work among researchers is increased according to the level of industrial application, notably patents. Indeed, following the work of Godin and Gingras (Godin and Gingras 2000), we know that the researchers who are most productive in publication activities are those who are also most involved in industrial partnerships; patenting is often preceded by intense publication activity (Azoulay *et al.* 2009), as this researcher explains:

> There is no technology development without high-level publication, they go hand in hand. There are people involved in technological developments for extraordinary vectors that have been published in *Nature*. Behind this is international visibility, recognition, everyone wants this product. Obviously, they go hand in hand! It's become essential that we become a key player on the international scene and that we publish high-level papers. We're all heading in that direction because, if we have high-level articles, we'll get grants and therefore money and that's our primary goal [...]. At my level, as a researcher, the goal is to publish at a high level, to patent, to bring in money, international recognition in meetings and so on... (Interview with Fabrice, researcher in an R&D laboratory accredited by the biocluster, January 2015).

In this interview extract, we see all the qualities of the star scientist identified by Caroline Lanciano-Morandat and Lise Gastaldi (Gastaldi and Lanciano-Morandat 2017): a large number of international publications, prestigious prizes, significant funding and so on. This researcher is in an R&D laboratory, so it is difficult to make him a textbook case, as researchers working in public laboratories are less subject to the patent issue. Nevertheless, his laboratory includes researchers from INSERM, the University of Evry and the CNRS. Socializing with the patent culture is certainly applicable to them too. Some authors see this publication–patent relationship as a system of cooperative exchange (Crespo and Dridi 2007) that allows researchers involved in industrial relations to access the resources necessary to maintain more fundamental activities. Others point instead to a destabilization of scientific work (Hubert and Louvel 2012, p. 20). The long time frames, supervised by peers, and the scientific requirements are constrained by the short time frame imposed by the industrial partner (Brunet 2012). Tenured researchers are therefore subject to a progressive acculturation of the research valorization model, through patents, which they then transmit to their PhD students. We can then observe sequencing strategies (Slaughter and Rhoades 2010) that are put in place, during the thesis, in order to reconcile publications with patents, as explained by this same R&D researcher:

> Here, there's a pretty big patent culture, so I tend to anticipate the patent very early on. For example, right now, I've applied internally for a patent. They assess whether it's of interest. If it is, then it becomes the subject of a PhD thesis, but there may be a delay of a year before the data becomes protected, so that means that you can't talk about it. So, it means the PhD student can't go to meetings, can't publish, etc. However, she will be in her last year of her thesis next year, so at that point, once the patent is filed, she can go to all the meetings and she can publish, but she'll have been thinking about it early on (Interview with Fabrice, researcher in an R&D laboratory accredited by the biocluster, January 2015).

This "portrait of the PhD student as entrepreneur" (Frances 2012) is particularly characteristic of PhD students whose thesis subject is applied research and who are therefore socialized very early on to this sequencing between publications and patents. Nevertheless, those who pursue more fundamental research also feel drawn towards businesses, as this bioinformatics PhD student explains:

> Well, that is to say that the most recognized research, and therefore the best funded, is the one that has the most industrial applications. I'm in basic research, which takes more time, so I publish less. If I don't publish, I'm not recognized. And if I do publish, I'm not fully recognized either, because it doesn't often come with a patent… I really regret it, but the job market in public research is so clogged up that I know the day will come when I'll have to sell my soul to the devil and work at L'Oréal (laughs). For now, I'm enjoying being able to do research… quote unquote (Interview with Anastasia, PhD student in bioinformatics in a labelled laboratory, September 2015).

This interview extract is consistent with the observation that basic research still makes sense to some researchers (Joliot 2013). The discourse on the end of an ideological barrier between basic and applied research contrasts with this feeling, perceptible in the PhD student's remarks, of the loss of legitimacy of basic research (Hubert and Louvel 2012, p. 24). Thus, we observe a heterogeneous relationship with industrial collaboration on the part of researchers. On the one hand, R&D laboratories participate in the construction of profiles that master the culture of both publication and patenting. On the other hand, basic research laboratories may host researchers who, conversely, have a scientific ethos in the Mertonian sense. This observation refers to the diversity of categories constructed by Lamy and Shinn (2006). Among researchers, the authors distinguish between three types of relationship to industrial logic: (1) the Academic, who attaches more importance to the scientific value of his work; (2) the Pioneer, on the contrary, who refers more to its industrial application. He does not give up on science, but he prefers to adapt it to economic imperatives. Finally, (3) the Janus, who can "signify his inclination for research without neglecting economic issues. Conversely, he can express his enthusiasm for entrepreneurship without adopting the same mercantile profile as the Pioneer" (*ibid.*, p. 38). Moreover, the call for industrial collaboration produces individual integration behaviors to varying degrees, between industry with science, without the dissolution of science into industry being completely achieved. Let us take a look at the cluster's business creators.

5.3. An unstable relationship between economic development and industrial secrets for companies

Start-ups are integral to innovation clusters, with the imaginary that surrounds them suggesting that creating a business is a deliberate choice for young researchers or students looking for an exciting adventure. The survey shows that, in reality, creating or working in a start-up is a more unstable process or organization than it seems.

5.3.1. *Individuals are increasingly encouraged to start a business to escape unemployment*

For the past 15 years, the switch to business creation has been taking place at an increasingly early stage. Not only are researchers encouraged to set up their own company based on their research, but students who have not yet entered the job market are also being encouraged to become entrepreneurs. Higher education has become a training offer, where students must choose not only courses, but also a pathway with the minimum cost and the maximum profit, to meet the needs of the labor market (Granger 2015, p. 11). The responsibility then falls on students to become actors in their educational pathways (Brucy 2012). They therefore become responsible for their own successes or failures in accessing the labor market. In addition to this responsibility in the choice of training, students are encouraged to create their own businesses.

The dissemination of this entrepreneurial spirit through schools is based on a set of European recommendations developed within a project that brought together experts from several countries, called "Best procedure project on Entrepreneurship Education and Training". These experts defined a series of desirable objectives for national education systems at all levels, from elementary school to higher education (Verzat 2015). This entrepreneurial directive can be observed in France as early as middle school (Tanguy 2016). Partnerships are set up between associations (the main one is "Entreprendre Pour Apprendre (*Learn by Undertaking*)") and the National Education ministry, so that middle and high school students create their own mini-enterprises. This diffusion is reinforced at the time of higher education. Business, management and engineering schools have been pioneers in this transmission, where lessons are oriented towards these individual entrepreneurship logics (Lamy 2017).

Universities are also increasingly training and encouraging entrepreneurship. If we take the case of the University of Evry, recently attached to the Paris-Saclay university grouping, we observe a series of devices, events or workshops proposed with this objective. For example, on March 13, 2018, the Life Sciences department of the University of Paris-Saclay organized a day entitled "*Sensibilisation à la création d'entreprises en sciences du vivant*" (Raising awareness of business creation in life sciences) in partnership with Genopole. The round tables were entitled "*De l'université vers l'entreprise: les ambitions de l'Université Paris-Saclay et Genopole*" (From university to business: the ambitions of the University of Paris-Saclay and Genopole) and even "*Témoignages de créateurs d'entreprise sur le plateau de Saclay et à Genopole*" (Testimonies of business creators on the Saclay stage and at Genopole). The University of Evry has also appointed an Entrepreneurship Officer and a Corporate Relations Officer, whose roles are to facilitate the introduction of students with the problems faced by entrepreneurs through a number of initiatives: the Entrepreneurs Café, Entrepreneurship Day, Workshops for prototyping business challenges and so on. The University of Paris-Saclay is also involved in spreading the Californian success story by organizing student entrepreneur trips to the heart of Silicon Valley. Since 2014 and the creation of the student entrepreneur status, these types of initiatives have been steadily increasing. This status is aimed at students in the course of their studies or young graduates who submit their business project during an application to one of the 29 Student Clusters for Innovation, Transfer and Entrepreneurship (*Pôles Etudiants Pour l'Innovation, le Transfert et l'Entreprenariat*, PÉPITE). Once selected, students receive support from a teacher and an external advisor and have access to the various PÉPITE networks. Student entrepreneurs can enroll for the student-entrepreneur (*Diplôme d' Etudiant-Entrepreneur*) diploma, which costs 500 euro per year, and can have their studies modified if their business project requires it. Genopole has been a pioneer in this area since 2005, 10 years before the creation of the student-entrepreneur status, when it hosted the IGEM competition:

> Every year, teams of degree students from all fields of science and from a variety of backgrounds spend their summers developing and realizing a project that involves creating biological systems with, as yet, non-existent functions from microorganisms and cells into which "bio bricks" provided by MIT are assembled (Häfner 2010, p. 986).

This time, the reference is not Silicon Valley, but another cluster that is also dominant in literature: that of Boston, centered around the Massachusetts Institute of Technology (MIT). This team highlights two dominant directives faced by students aiming for a career in research: on the one hand, to professionalize in the development of projects with high technological potential during their studies and, on the other hand, to position themselves in globalized competition.

In general, business creation is seen as a way to address youth unemployment, which has been steadily increasing since the 1980s (Le Rudulier 2016; Toutain and Verzat 2017). Indeed, the "auto-entrepreneur" or student entrepreneur statuses are presented as a way to escape the constraints of big business (or academia) and to respond to liberal values of freedom, self-actualization and succeeding in private enterprise. In reality, behind these statuses, insecurity, that is, unemployed or young graduates without income (Abdelnour 2014), is mostly hidden, even though official speeches claim to reduce unemployment through these same devices.

With its Shaker program, the biocluster is fully involved in encouraging young graduates and students to create their own businesses. In addition to the classic accompaniment of researchers supported by Genopole business managers in the realization of business plans and fundraising, the cluster wishes to intervene as soon as possible with future entrepreneurs:

> We are going to launch a Shaker project aimed at young people who want to create a business directly from the results of their science thesis. Knowing that they will no longer be going to INSERM or CNRS, these researchers are – thinking: "Why not make my thesis serve two purposes? On the one hand, a very important degree in science and, on the other hand, a start in life to develop a business project". With this Shaker, we are going to develop a support and hosting arrangement for these very young researchers. It is interesting to see that this mobilization movement is developing in a somewhat paradoxical way, because the public sector is no longer able to offer positions in sufficient numbers (Interview with the general manager of the Biotech *Finances* biocluster, No. 754, January 30, 2017, p. 3).

Literature has identified two types of entrepreneurship: those motivated by "pull" factors, which refer to the challenge, the need for recognition, the availability of money, the desire for independence and so on, and those resulting from "push" factors, such as escaping from unemployment, coming back from a layoff or simply escaping a professional situation deemed unsatisfactory (Nakara and Fayolle 2012). The creation of start-ups is encouraged by institutions, such as the OECD and Genopole, in order to respond to a "push" factor, that being the high unemployment rate among young graduates. However, although students and young graduates are particularly encouraged to create their own businesses, they are far from being representative of the cluster's entrepreneurs. Moreover, the logic outlined here, which is eminently individualistic, also seems difficult to reconcile with the claims of collaboration.

5.3.2. *The fragility of the male 30-something entrepreneur*

Management literature abounds on the dithyrambic description of entrepreneurs as a people outside the norm: they take risks, are autonomous, adapt quickly and know how to mobilize a network (Buquet *et al.* 2017,

p. 97). They have all the characteristics of the heroic figure described by Schumpeter, who breaks the economic routine through a process of creative destruction (Schumpeter 1934). "Lifestyle" entrepreneurs (Gomez-Breysse 2016) are described as resistant to stress, which makes them high performers. Unlike executives, who would face a difficult reconciliation between personal and professional life, entrepreneurs are in harmony with their professional choices and view business creation as a self-accomplishment. This vision echoes the works that describe the figure of the artist as the archetype of the modern worker who is hyperflexible, autonomous, creative, capable of acting in uncertainty, who advocates freedom and who mocks the hierarchical and organizational constraints of the traditional enterprise (Chiapello 1998; Menger 2003). Chiapello sees a coming together of the logics of art and management, while Menger postulates that the characteristics of artists prefigure those that the world of work requires from now on of workers as a whole: "hyperindividualism, creativity, mobility, taste for risk and novelty, acceptance of forced vacations and of the most blatant inequalities" (Rameaux and Ribac 2003, p. 146).

This imaginary of the bold visionary was forged around a few personalities who put their whole being in the service of their company and/or brand. Through presentations (pitches) in front of full houses, they put forward the extraordinary potential of their project. One thinks of Steve Jobs' highly publicized presentations to promote the latest technological products marketed by the Apple brand. Silicon Valley is particularly well known for the figure of the entrepreneurial "winner" who emerges in the context of a technology cluster: a young American man in his thirties or still a student, most often working in the electronics and information technology fields, with a technologically revolutionary project (Barthe *et al.* 2016). Steve Jobs at Apple, Bill Gates at Microsoft and Marc Zuckerberg at Facebook are the best-known examples.

However, beyond speeches and representations, several studies show that the daily life of these entrepreneurs is not limited to this embellished vision of business creation. In the same way that the success stories of the Silicon Valley and MIT clusters are very much in the minority compared to the failures that mark the many initiatives that are taken there, the few successes of entrepreneurs are generally highlighted, whereas, more often than not, they experience failures, financial setbacks and abandonments that can turn into disasters. This is what Grossetti, Barthe and Chauvac show in a survey conducted between 2005 and 2014 on 97 innovation companies located in

Toulouse, Castres, Tarbes, Bordeaux, Grenoble and Marseille, whose founders' daily lives they dissect:

> The entrepreneur is not a particular type of person, it is only a social figure constructed by references shared in varying degrees and devices that encourage, or even compel, those who wish to create businesses to conform to the expectations that are associated with this figure (Grossetti *et al.* 2016, p. 91).

By reconstructing their stories of business creation, they show that their paths differ from the famous figures of the start-up world. However, they share a common characteristic: 93% of them are men (*ibid.*, p. 168). At Genopole, the observation is the same: in 2016, of the 75 company directors, only 14 were women (81% male directors). This large majority of men among entrepreneurs seem to be the only characteristic common to the aforementioned archetype of the 30-year-old student or young graduate. In fact, again in 2016, there were few 30-year-olds among business leaders. 62% of men who run a business are over 50, less than 2% were under 30 and 15% were between 30 and 40. As for women, the few who start their own businesses do so mostly after the age of 50.

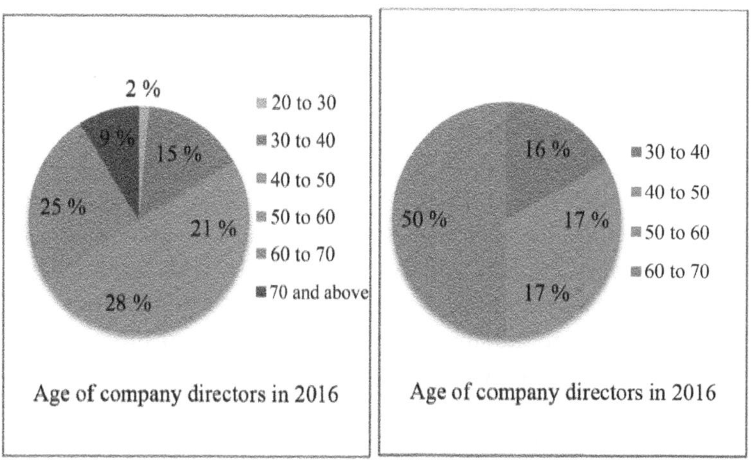

Figure 5.1. *Age of company directors (source: graph constructed from the Genopole database, based on 75 companies accredited by Genopole in 2016, domiciled in Evry and the Paris region). For a color version of this figure, see www.iste.co.uk/vallier/clusters.zip*

The vast majority of these are experienced researchers, over 50 years old, who, after a career in public research, have developed a technology that can be converted into a business. A minority of young researchers have set up their companies from their thesis or post-PhD works; in 2016, there was just one. On the contrary, among the youngest founders, we find more engineers or business school graduates, who have most often set up their businesses with others.

Thus, whether it is based on a network of knowledge acquired over the course of a career, or whether it is set up by several people, the creation of a business turns out to be a collective activity, made up of ordinary innovations, far from the image of the young, solitary and heroic entrepreneur (Alter 2000). In this respect, the entrepreneurs all relied on the cluster or other public support to get started. The role of an intermediation structure like Genopole is to provide project leaders with skills outside their scientific training (fundraising, recruitment, network development, etc.):

> After a career of about thirty years, I'd had enough of contract work, training students and then seeing them leave because we couldn't recruit them… I was tempted to go into the private sector, I tried to get recruited in industry, but when you're 55, it's complicated, it didn't work out, so I set up my own business while staying in my lab. The first year […] yes, the support from Genopole went well, at first, even very well, after moments of doubt: they didn't believe in it at all, it was really impossible for me to succeed in setting up a company… I'm not someone much of a show-off, as it were, and so they didn't see me succeeding, but then that changed. I had a business manager who was very present and helped me a lot with applying for public assistance and the departmental competitions. They also provided managerial support by taking on part of the cost of a business developer, which was very valuable at the beginning (Interview with Sylvie, director of a company accredited by the biocluster, June 2015).

This former INSERM researcher's comments show that Genopole's business managers did not see her as the ideal profile for a start-up founder: a woman researcher over 50. Resources were therefore concentrated on helping her develop the skills needed to create a business. Moreover, we observe that her desire to create a company, based on her research, is the result of professional dissatisfaction with public research, particularly the procedures and the lack of recruitment. In this regard, some observers see researcher entrepreneurship as a way out of their professional difficulties, rather than as a way of improving research (Heil 2010, p. 376). The study conducted among start-up founders comes to the same conclusion: the creation of an innovation enterprise is rarely the result of risk-taking, but more a solution in an unstable context (Barthe *et al.* 2016). Here, the survey highlights almost contradictory profiles among entrepreneurs. Indeed, there are rare cases that honor the figure of the researcher entrepreneur, at the intersection of science and industry, mastering both scientific and entrepreneurial skills. In the majority of cases, as we will see in more detail, they are "entrepreneurs of circumstance" (*ibid.*), some of whom may even have difficulty in earning a salary.

5.3.3. *Intense activity marked by the search for financing and competition*

The definition of a start-up is similar to that of an innovation company, in terms of its sector of activity, its marketing methods or its mode of development, but it differs from it in that it grows rapidly, in terms of revenue and capital (Hurel 1999, p. 6). Some of the firms on the site are in this fast-growth configuration, in terms of sales and staff, but they remain a minority. Most of them have difficulty developing their activity. In this context, entrepreneurs rely on the cluster, or other public support, which may become essential to the survival of their business. Let us listen again to the former INSERM researcher, now a company director, who reports on her financial situation vis-à-vis her former research colleagues:

> I don't think they realize the financial difficulty that we find ourselves in, I don't think they know… Me, when I explain to them that, every month, I have to find 1,000 euro of salary (laughs), it's another world… (Interview with Sylvie, director of a company accredited by the biocluster, June 2015).

We recall that this director of a company of five people has, among her employees, a CIFRE PhD student who is doing her thesis with an accredited laboratory. She was referring to the divide that exists between a laboratory that welcomes a PhD student in order to develop its research policy and the necessary, even vital, commercialization of the results of the thesis for the survival of the business. This fear of failure can then cause damage. The founder of a start-up is very often identified with his company. Thus, keeping the company afloat or witnessing its failure can create a lot of anxiety. We recall that, between 1998 and 2016, 29 out of 180 accredited companies ceased trading, 9 underwent a buyout or a merger and 56 left the biocluster without knowing if they still existed. In this regard, let us listen to a company director on the consequences that a merger can have:

> In 2005, our Japanese investors, inbetween raising funds, merged with a big pharma. But this big pharma was not interested in cell therapy products in the broad sense. So they didn't insist in the board meeting and stopped our activity, even though the product was active... In the United States, it took a fortnight, in France it took three months. At that time, we had 60 people in the US and 40 in France. So, I found myself with a lot of debt [...], I had to manage as I best I could. Apart from laying off all the staff, I had to negotiate procedures against research agreements [...]. So, we ended up finding a buyer for the lease and the equipment, which allowed us to transform a desperate situation into a manageable one, a reasonably clean and neat closure (Interview with Alexandre, director of a company accredited by the biocluster, September 2015).

As we can see here, healthcare start-ups are highly susceptible to being bought out by a big pharma. In the interview extract above, we observe that the takeover can have disastrous consequences for the company: cessation of activity, laying off of staff, debt and so on. In this uncertain context, a recent study shows that entrepreneurs are highly prone to depression: 30% of them compared to 75% of the French population (Buquet *et al.* 2017). The involvement of these individuals can be limitless, while the payback is often small. As the company director above explains, she struggles to earn 1,000 euro in salary per month. Moreover, even though some manage to find a financial balance, highly active businesses with more than 10 employees are few and far between. Indeed, start-up founders and their employees are now known for their high level of involvement:

> Each person is fully committed because they are fully convinced, which abolishes any need for hierarchy and makes the delimitation of working hours and the obligation to pay a salary secondary; the start-up brings together individuals driven

> by the same desire to give life and future to the project of the creative leader (*ibid.*, p. 105).

The organization of work in a start-up context is therefore very intense for both the founders and the employees who work there. In addition, we recall that 73% of the biocluster's industrial fabric is made up of start-ups with fewer than 10 employees. In this context, the division of labor is not as clear as in a traditional business; each person may have to perform several different tasks, leading to over-solicitation. In addition, there are cases where this multi-tasking can be carried out by one or two people. In fact, the survey counted about 10 companies where there were fewer than two employees. Their working conditions can be very lonely, as recounted by this biomechanics engineer, who works alone with her manager and who is often on the move looking for investment

> I have to spend days where I don't see anyone, I eat by myself, I go back to my office and in the evening I leave. Sometimes I say to myself, no one would see if I came in or not. And that? It was quite a shock. Afterwards, the director tried to be present at the start, but after a while, he'd often be in meetings. He's also looking for investors, business plans, and so on, so he's often away. I can't wait for other people to come into the company so that I can really have professional relationships, even friendships (Interview with Rizlaine, accredited company biomechanics engineer, December 2014).

Moreover, the fields of activity of these start-ups and the projects they develop, however diverse, are all presented as ways to change the world. The employees within these companies are all the more involved as they are convinced to participate in solving issues related to genetic diseases, agriculture or even food:

> To finance ourselves, we did friends and family, we are on the fourth time. We've raised a million euro in four rounds from private investors, but very private: friends and relatives. And we still can't convince venture capitalists, because there is a very strong change, that is, there is a total financialization of the thing. At the start, we said, here we are, we have new tools to look at the world, and here is the evidence. There is some magnificent scientific literature. But, no, there is no interest, so

> we say, we will sell products, we work on a diagnosis [...] for children we know how to treat but we don't know how to diagnose. So we say, "our tool is really great!" (Interview with Michel, director of company accredited by the biocluster, December 2014).

This interview extract highlights just how convinced this entrepreneur is of the nobility of his project, even though it is not financed. It also shows how far the involvement can go: resorting to borrowing from family and friends. This commitment to changing the world, coupled with the small size of the Genopole companies, creates an intense workload with high hourly rates (evening and weekend work, little time for lunch breaks) and requires versatile workers.

Finally, the competitive context for start-ups in the midst of process development may force employees to maintain a certain confidentiality and provoke strategies of concealment, which may affect their sociability. We recall that, out of a sample of 25 accredited companies, only 5 have formal relationships with each other (see Chapter 6). In addition, 45% of the individuals who took the questionnaire stated that they were bound by confidentiality with regard to other cluster structures and 60% of them worked in a company. This confidentiality requirement is not specific to VSEs, but is also found among SMEs with between 30 and 40 employees. They represent 25% of the cluster's industrial fabric and, within the Evry area, they do not exceed 60 people. Some of these SMEs are still in the research and development phase of a process to become market leaders. During the survey, the management of one SME expressed a strong reluctance to fill out the questionnaire for the network analysis. However, the questionnaire was administered in a one-on-one interview with the CFO, who then warned of "management's fear that what was said in the questionnaire would be made public". During this interview, it was explained that the company's employees are subject to a confidentiality policy. Thus, each new employee is obliged to sign a document that commits him or her not to reveal any information. The document specifically states that they are not to talk about it in areas shared with other biocluster structures (intercompany restaurant, coffee machines, etc.). The CFO justified this caution by the fact that at the time of the interview, the company was in a publication phase around their technology in order to obtain European funding. Not revealing their research is a way to avoid cutting themselves off from future financial resources. Hence, in order to achieve realization of their process and be the first on the market, the CFO explained that the employees follow two rules, confidentiality and time optimization, which leave no room for external sociability. These two elements are the main factors that prevent individuals in the cluster from participating in networking (see Chapter 6).

6

The Avoided Cooperation

The two structuring social fields of the cluster are spanned by internal struggles that limit exchanges between accredited members. This final chapter then focuses on the form of the network produced by these strained fields. Beyond the words of the respondents or the sponsors who testify to a lack of cooperation, carefully observing and categorizing the nature and volume of relations in the cluster makes it possible to realize the extent to which cooperation is finally avoided.

6.1. A patchy local network

While the local shape of the network is important in understanding the types of interactions that take place within the cluster, the importance of external relations also shows the problems posed by geographical proximity.

6.1.1. *What type of organizations for what type of interactions?*

Using the network analysis method, relationships within the cluster could be mapped from a significant sample. Many examples in network literature are based on researching groups with defined boundaries (school classes, offices, etc.). In contrast to observing a network in a business with well-defined boundaries, the cluster poses problems in identifying boundaries, as it involves sectoral and/or territorial delimitations (Bernela and Levy 2017, p. 5). Hence, in the context of a perimeter marked by a diversity of subjective imprints that blur its lines, the choice has been made to rely, on the one hand, on its institutional contours, constructed and conveyed by the Genopole structure, and, on the other hand, on the unity of place represented by the Evry region. The intermediation structure recognizes the fact that companies and laboratories belong to the biocluster in three distinct ways: (1) by accrediting them; (2) by listing them in the annual biocluster directory; and (3) by communicating with them institutionally through promotional events and brochures. As a result, the network includes 56 companies (41 VSEs, 14 SMEs and a private organization providing work experience training in biotechnology professions), 20 laboratories, 13 of which are attached to the local university, two private laboratories, a patients association, the hospital and a scientific culture association.

Structure attributes	Corpus of data	%	Sample	%
Patients association	1	1%	0	0%
Scientific culture association	1	1%	1	2%
Hospital	1	1%	1	2%
Small private laboratory (< 50 people)	2	3%	2	5%
Small public laboratory (< 50 people)	9	11%	4	10%
Medium-sized public laboratory (50–150 people)	7	9%	6	14%
Large public laboratory (150–300 people)	2	3%	2	5%
Private training organization	1	1%	1	2%
SME	14	18%	8	19%
VSE	41	52%	17	41%
Total	**79**	**100%**	**42**	**100%**

Table 6.1. *Typology of network structures*

In Table 6.1, the “Corpus of data” column shows the population to be observed, broken down by structure type and the percentage represented by each. In total, the whole network comprises 79 structures, 42 of which make up the sample. The sampling method is similar to that used in the analysis of the innovation environment of the Jura Arc, whose results are based on a survey of a sample of businesses (Pfister and Nemeti 1995, p. 32). The different structures are categorized according to their size and status. From previous studies, the nature of partnerships can be measured by, for example, separating SMEs from large firms (Levy and Talbot 2015), but in order to adapt to the scale of the Genopole network, VSEs or start-ups have been dissociated from SMEs (who do not exceed 60 employees in the area studied). Finally, it should be noted that the intermediation structure does not appear in the list of structures for the network studied. Indeed, by accrediting and supporting laboratories and companies, the intermediation structure is in contact with all the members of the cluster. To include it in the analysis would have led to a star-shaped graphical representation that would have overshadowed the object of our attention: the relations between accredited members[1].

While network analysis allows for the mapping of relationships, it is the interviews and observations that allow for their nature to be known. Indeed, the categorization of relationships between cluster members was developed during interviews and observations, which are an essential step in "discovering the nature of structuring social resources likely to be exchanged between actors" (Eloire *et al.* 2012, p. 81). In the literature, there are a large number of relationship typologies, according to the authors. Economist Nicolas Carayol outlines five types of collaboration (Carayol 2003): (1) a laboratory performs technological service work on behalf of the company (whose financial outlay and risk-taking are low); (2) a bilateral partnership in which industrialists can hire students and postgraduates, and academics can benefit from subsidies as well as targets (businesses) for their research; (3) the investment of large groups in fundamental research, where the potential for technological disruption is considered to be very high; (4) more modest collaborations in terms of scientific and technological potential, but which, nevertheless, allow companies to receive scientific recognition and to develop innovative technologies; (5) the final relationship typology refers to important research consortia in which companies and

1 The data from the questionnaire show that only one manager declares that he has not had any dealings with the intermediation structure since his appointment. On the contrary, this is not the case for the other managers, 25 of whom even state that they have been in contact with the general management of the biocluster. In addition, the intermediation structure is in contact with the structure group as a meta-organization and resource actor (economic support, fundraising, communication, research support, pooling and supply of equipment, etc.), which does not correspond to the relations exchanged in the rest of the cluster.

laboratories benefit from significant public subsidies in order to create research networks with shared results. Perkmann and Walsh reduce the typology to three categories of collaboration (Perkmann and Walsh 2007): research services (industrialists call on researchers and pay them for expertise, technological and/or scientific competence), research partnerships (formal collaborations around research and development activities) and collaborative research, bringing together researchers and industrialists around a common research problem.

As we can see, these different exchanges are not dependent on spatial proximity. Nevertheless, empirical studies agree on the existence of at least partially non-market exchanges between organizations belonging to local systems, which may take the form of exchanges of information, technical assistance, peer support and so on (Grossetti 2004, p. 164). Drawing both on existing categories in the literature on relations between science and industry (Estades *et al.* 1996; Carayol 2003; Grossetti and Bès 2003; Perkmann and Walsh 2007; Renaud 2015) and adapting to the particularities of the field studied, the study of interactions at Genopole draws up a typology presented in Table 6.2.

Type of interactions tested	Essential characteristics of the interaction
Collaboration contract	Contract that formalizes an agreement between two structures to pool resources and knowledge with shared results.
Consortium	Research and innovation project whose financiers (competitiveness cluster in France or principally the EU's H2020 program) require the association of industrial actors (companies) and scientists (laboratories, research organizations, universities, etc.).
Subcontracting	Contract by which one company requests another to carry out a part of its production.
Customer-supplier	The sale of a product or process by a supplier to a customer who purchases its goods or service.
Cosupervision of thesis	Supervision of a PhD student by two people belonging to two different institutions, between two laboratories or between a laboratory and a company (CIFRE).
Assistance relationship	Loan of equipment or technical assistance without commercial exchange.

Table 6.2. *Typology of cluster interactions*

6.1.2. *Scientific and market relations behind informal interindividual exchanges*

Both the structure typology and the relationship typology have been integrated into the Ucinet software package (see Figure 6.1). Relationships involving reciprocity are illustrated by a double-headed arrow: these are collaborations, consortia and thesis co-directions. For subcontracting, help desk and client–supplier relationships, the arrow indicates the direction of the relationship[2]. It should also be noted that, in some cases, two network members share more than one type of relationship. In these situations, for the sake of clarity, a hierarchy has been applied that consists of prioritizing the relationships in the following order: collaboration contract, consortium, subcontracting relationship, client–supplier, thesis cosupervision and assistance[3].

The interorganizational network thus obtained offers a representation of the tensions that operate in the cluster, while at the same time providing accurate information on the links maintained by some of the members. We can see that collaboration contracts structure most of the exchanges. Laboratories are the most involved in this type of interaction, equally with companies (11 dyads[4]) and also among themselves (11 dyads). The predominance of laboratories in this type of relationship can be explained by the fact that research is increasingly structured by contracts with other laboratories or businesses that enable it to finance its work or to have access to resources, such as research teams or equipment (Cassier 1996). Relationships between laboratories are not limited to collaboration contracts and we observe that they are also modestly linked by consortiums and thesis codirections. The laboratories are also in relationships with a third type of actor, which was not included in the initial sample: the hospital. In the case of the hospital, its position within several triads can be explained by its ability to make cohorts of patients available to businesses specializing in health for the clinical testing phases.

---

2 In the graph, the direction of the arrow from X to Y indicates that X subcontracts to Y, that Y is X's supplier, or that X troubleshoots Y.

3 Without going back over Porter's definition in detail, the concern of clusters is to develop formal collaborations, whether commercial and/or scientific, in the innovation process. Thus, contractual relationships appear as a priority, before tacit exchanges. As a result, troubleshooting relationships are much more important than they appear in Figure 6.1.

4 Dyad: link between two actors of the network.

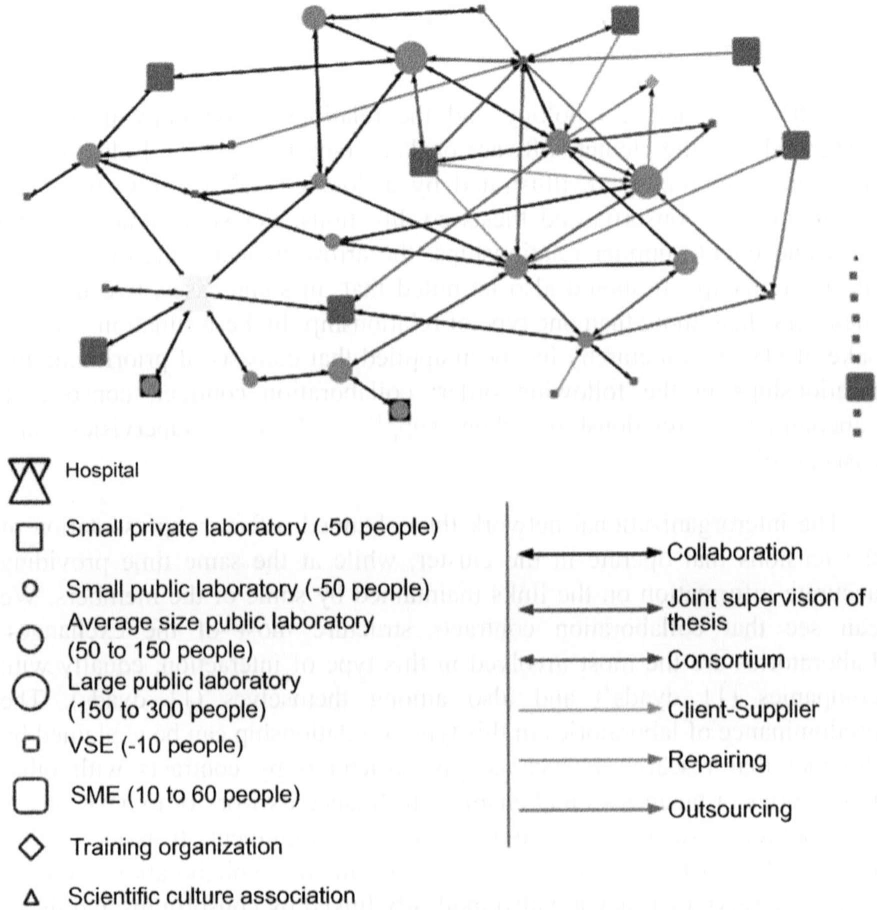

Figure 6.1. *Network between 42 biocluster structures[5]. For a color version of this figure, see www.iste.co.uk/vallier/clusters.zip*

Relationships between industrialists are almost non-existent, mainly due to competition and industrial espionage, which is one of the negative externalities of clusters highlighted in the literature (Torre 2006; Depret and Hamdouch 2009). As for relations between science and industry, we observe that laboratories and companies are relatively linked by the different types of

5 For a more detailed analysis, see Vallier, E. (2018). A journey in cluster: what about its relational promise? *ARCS – Analyse de Réseaux pour les Sciences Sociales (Social Sciences Network Analysis), GDR Analyse de réseaux en SHS (Human and Social Science Network Analysis Research Group).*

scientific and commercial relations: collaboration, consortium, subcontracting and so on. An additional form of interaction between science and industry, which escapes network analysis, nevertheless emerges: the creation of companies from technologies developed in laboratories. Unlike an exogenous relationship with another cluster member, this is an endogenous process, known in innovation jargon as a "spin-off". It refers to the carrying forward of an innovation or a scientific process developed within a structure by the creation of a company. 19% of the structures participating in the questionnaire were created by another structure in the cluster and 26% originated from new companies.

Customer–supplier relations appear particularly patchy, given that one of the three externalities of the Marshall district is the emergence of specialized suppliers (Branciard 2002), as well as subcontractors, who form the pillars of the Porterian cluster, along with producers and customers (Leducq and Lusso 2011). The methodological choice of showing "formal" relationships first on the figure is a reminder of the need for graphical clarity. As a result, informal assistance service relationships (loan of equipment, technical assistance, exchange of information without commercial exchange and/or contractualization) are under-represented. In fact, the figure shows eight such relationships (in blue), whereas there are 14 in total. This type of relationship therefore comes second to collaboration contracts.

These assistance relationships observed on an organizational scale were also highlighted in the relationships between individuals working in the cluster. Indeed, 27% declare having helped another individual in the cluster and 19% declare having been helped by another person in the cluster:

> At the production level, from the labs, we are starting to know our neighbors better. When we need something, we can go knock on their doors. We still have access to things, but it's limited to the loan of small equipment (Interview with Emma, accredited company research engineer, July 2015).

Geographical and technological proximity (using the same equipment and facing similar problems) leads to "knocking on neighbors' doors", provided that they know them beforehand, as the interview extract above underlines. This spirit of assistance is similar to informal links such as bartering or material exchanges, highlighted by Faulkner and Senker (1994), but does not constitute a perennial reciprocal interaction insofar as "mutual aid and cooperation cannot be confused with a succession of 'helping hands'" (Alter 2002, p. 269). Interindividual exchanges are therefore convivial[6], even friendly[7], but rarely convert into professional collaborations. Moreover, most of these links seem to have been forged in the relationships of former students or colleagues, rather than in brief encounters on the cluster. The somewhat outdated and colorful expression "cafeteria effect" (Lipinski 2013), which consists of thinking that, by working in close proximity, individuals with different institutional affiliations end up setting up collaborative projects from their informal meetings around a coffee machine or restaurant table, has its limitations. As Claude Fisher has shown, most individuals meet through structures that are considered stable and enduring (family, work, neighborhood and school), while brief encounters in a bar or restaurant rarely convert into strong ties (Fischer 2011).

On the contrary, the fact of having shared the same studies or the same employer demonstrates a professional circulation within the cluster. Nearly a quarter of respondents (23%) say they have worked in at least two structures in the biocluster. This local job market is particularly encouraged by laboratory directors and company managers who, according to the interviews, form the "little world of genomics". Here again, the interknowledge of managers appears to be used more to identify and capture skills than to establish collaborations. These collaborations seem to take place more outside the cluster.

6.1.3. *More outwardly looking organizations*

Of the 32 managers questioned, 57% put French institutions not based in Genopole at the top of their list of cooperative ventures, 28% said foreign institutions and 15% said institutions based in the cluster (see Figure 6.2).

Having focused attention in the previous chapters on exchanges at the local level, it is now appropriate to ask the question of the insertion of actors in networks outside the biocluster. The literature (Bathelt *et al.* 2009) shows the strategic importance of external links (global information pipelines) in the development of a cluster without dense internal networks ("local buzz").

---

6 55% of respondents report having developed friendly relationships with people from cluster institutions.

7 Of the individuals surveyed, 23% admit to having friendships across the cluster.

Moreover, the biotechnology market has a strong transnational dimension that has largely developed in the industrialized countries of Western Europe and North America (Foyer 2006). Moreover, biotechnology fits the research and innovation model advocated by the OECD within its member countries. It is therefore not surprising to find that 58% of the sampled organizations' trade is with Western Europe and the United States[8]. In the end, several authors question the legitimacy of the local scale, at the expense of national and international relations, as a relevant framework for analyzing innovation activities in an increasingly globalized context (Boschma and Frenken 2009). This point of view is echoed in the population surveyed at Genopole, as in the case of this teacher–researcher:

> Is geographical proximity an advantage? Theoretically, yes, but today communication methods have evolved enormously, researchers are used to traveling and finally it's no easier to collaborate with someone who is 500 meters away than someone who is 500,000 km away (Interview with Emmanuel, teacher-researcher and laboratory platform manager, July 2014).

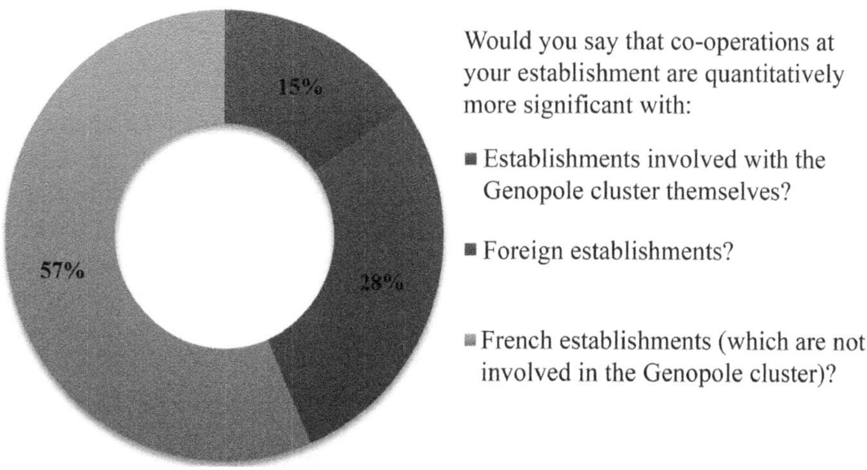

Figure 6.2. *Intra- and extra-cluster organizational relationships (in France and abroad). For a color version of this figure, see www.iste.co.uk/vallier/clusters.zip*

8 In fact, in the questionnaire, while company managers and laboratory directors answered that their relations with foreign countries were quantitatively more important, the questionnaire was designed to ask them about the geographical area of these international exchanges.

The idea contained in this extract, where geographically distant actors may be able to maintain relationships through physical (he mentions the habit of traveling) or virtual (new communication technologies) exchanges, is in line with theoretical findings (Hamdouch and Depret 2009, p. 13). This trend is particularly strong in the biopharmaceutical sector, where links with large national and international industrial groups are sought after (Owen-Smith and Powell 2004), which is corroborated by this comment made by a company director of studies:

> In fact, our customers are more big pharma or middle pharma, but internationally, in fact, a little bit nationally but not much. Anticancer molecules is not really their field, either that or they are not advanced enough to be able to do it. We don't have any clientele here […] and, in terms of collaborations, we do have some, but not here. We do have some with the Curie Institute, with Cochin, with a hospital in Marseilles and in Montpellier […]. The company was created with the Curie Institute and it provides us with tumor models […]. After that, we have collaborations in the United States, but this is more for joint development, at the technical level (Interview with Leïla, accredited company director of studies, July 2015).

In addition to big pharma overseas, this respondent's company has collaborations with scientific institutes or hospitals, but which are located elsewhere in France. The same observation can be found on a quantitative scale in a 1996 study which reveals that 70% of the formal links between academics and American biotechnology companies are made outside the local framework (Audretsch and Stephan 1996). In a context of geographic clustering of innovation activities, how can we explain that the local level does not appear to be the relevant level?

6.2. Cooperation prevented by paradoxical demands

This second part of the book has repeatedly highlighted the existing tension between the individual logics that run through the organizations and the cluster's desire for cooperation, which is effectively perceived and experienced by the respondents as taking second place to an already busy job.

6.2.1. *Additional time pressure*

The different forms of networking all have in common that they mobilize the workers of the site outside of working hours of their company or laboratory. However, their working time, whatever their organization, already seems saturated. Indeed, workers in laboratories are increasingly subject to constraints in terms of workload. For a long time, researchers' working time has included a diversity of activities between research, teaching and valorization. The so-called administrative activity is also becoming increasingly important, around stewardship and university management tasks (Dahan and Mangematin 2010). These tasks are linked to the 2009 legislation on university autonomy, and also to the logic of calls for projects that require administrative skills (see Chapter 5). Peripheral activities (procedures, reports, minutes, funding applications, etc.) are gradually nibbling away at the researcher's core business (Hubert and Louvel 2012, p. 22). Faced with this plurality of activities, researchers' working time is closer to that of the self-employed or the liberal professions (not counting time, evening and weekend work, advancing research during vacations, etc.) than to that of the administrative staff, who support teaching and research activities (Gastaldi and Lanciano-Morandat 2017).

This strong commitment to work is all the more significant because it is internalized as voluntary and mandatory participation from the postgradutate, which constitutes, from this point of view, a long period of socialization, with a "need to work a lot, which ends up being 'naturalized' as evidence, a norm, which is then no longer questioned" (*ibid.*, p. 4). With tenured staff, the expected time availability can even be almost 100% among young researchers (Del Río Carral and Fusulie 2013), who see their involvement as a way to stabilize their non-tenured status within laboratories. The "I am overwhelmed" of the teacher–researcher (Aït Ali and Rouch 2013), highlighted in the literature, is accompanied by frequent displacements, linked to the spread of university buildings in the city of Evry. Let us listen to this teacher–researcher in this regard:

> We're still on a very fragmented campus and, as we have classes, we are already walking around quite a bit. We give courses in several buildings in Evry, all at a fairly fast pace. So, when you've already gone from one to another… well, there you go, you can't do everything… (Interview with Blandine, university laboratory director, October 2015).

As we have seen previously, time optimization is also a must in companies that then neglect the network dimension, as this company manager explains:

> The problem is that I'm often caught up in a lot of things. When I know in advance, it's good. We have very time-consuming activities during the day, I have a lot of things to do, I constantly have to talk with the United States, talk with my partners, etc. I don't have to communicate to promote the company, I have to communicate with my partners. I don't have any communication to do to make the company known, I have my own internal management to manage, the networking dimension takes a back seat (Interview with Philippe, leader of a company accredited by the biocluster, December 2014).

Whatever their organization, individuals are asked to participate in cluster activities. However, for most of them, this means concentrating on the internal activities of their respective organizations, which represent a significant workload. In this general context of reduced availability, networking creates additional time pressure. The workers agree to make use of their lunch breaks to participate in cluster activities. A consultation carried out on the expectations in terms of cluster animation indeed points out this phenomenon. In anticipation of the opening of the multi-purpose room, L'Escale, a question was asked about the preferred times to use this room. 37% of people declared that they would often or fairly often use the room between 12 noon and 2 pm. However, the short duration of the lunch break (75% of the 534 respondents take between 45 minutes and 1 hour 15 minutes for lunch) defines a greatly reduced availability, even though this is the preferred time slot. The end of the day, on the contrary, is reserved for prolonging work or going home:

> Genopole has always had these very commendable efforts to have a campus life. There are picnics, I feel like it's getting a little bit better, but hey (sighs). Evry is still a town where you don't stay. There are a lot of people who work here, but who don't live here, starting with me. I live in Paris itself and, after work, you prefer to go home. At Harvard and Cambridge it's a little different. This campus life really is something very American. We've tried to copy and paste that model here, but it

> doesn't work too well... (Interview with Patrick, research institute director, September 2015).

In this case, the desire to go back home is reinforced by the fact that Evry is not a place where people want to stay. In addition, there is recognition of efforts led by the biocluster, but whose copy-paste of the North American model is not appropriate to the local configuration. Here, the observation is far from following Dujarier's idea that "working with programs implies adopting their rhythm [...]. Programs, in fact impose a uniform and universal cadence of their own" (Dujarier 2015, p. 37). With the cluster, we observe *a contrario* that individuals have no interest in adapting their personal rhythms to that of the "campus life" mentioned in the interview above. This feeling that the biocluster's strategies are not adapted to the individual and collective practices of the site's organizations is quite widely shared.

6.2.2. *A disembodied objective between prescribed program and real work*

The additional time pressure exerted by the cluster produces a strong distortion between the discourses it promotes and employee practices (Linhart 2010). Indeed, the ambition to modernize programs does not radically change organizations overnight. Moreover, the ambition to "get managers at different levels and the employees themselves to sign up to a vision of the world created from scratch" (Linhart 2015, p. 48) cannot be achieved simultaneously and homogeneously. This discrepancy between discourse and practice, and between the practices themselves, produces asynchronous logics. Depending on the hierarchical levels or departments, new programs will not be integrated in the same way or within the same timeframe by individuals:

> What I see is that Genopole is developing many tools to help people interact. After that, does it work or not? In my small way, in my field of applied research, I'm still in the decoding phase. So, as a result, we are not necessarily in the same dynamic. Start-up dynamics are based on a goal and a schedule, so we also work with that, but it is perhaps less constrained because we don't have the commercial side onside. So, of course, we don't have the same dynamic (Interview with Blandine, university laboratory director, October 2015).

This laboratory director says she is well aware of the "many tools" developed by the biocluster, but, in her opinion, the objective of interaction with a company seems to be disembodied from her fundamental research activity. Moreover, we note the dissonance between scientific and commercial temporality mentioned at the start of this chapter, where the program is known about, but it does not meet with interest. In her book, Dujarier refers to the program designers as "planners", an appellation which, as well as referring to their function of creating plans, also refers to a characteristic that they often begrudge: that of "planing", or hovering, over physical work situations. Several of the people interviewed spoke of the secondary aspect of participation in activities in the face of the managerial and financial imperatives to which accredited organizations are subjected:

Genopole cannot hire people to do this work. We need volunteers because, realistically, the money would be better spent elsewhere. Especially those of us in science, we need money every day, we'd rather buy equipment than have Genopole hiring three clowns to do animation… (Interview with Marine, accredited laboratory study engineer, September 2014).

In this extract, the need to be "realistic" from a financial point of view is confined to the recruitment of "three clowns" to ensure the animation of the cluster. Whereas, in Dujarier's analysis, "planners" accuse opponents of the programs of being "dreamers" (Dujarier 2015, p. 67), we note that here it is the intermediation structure that is blamed for its lack of pragmatism. For this commercial manager, it is important to "talk about sales figures" or "export assistance". She even feels that the proposals for collaboration show that Genopole has not understood her imperatives. In a few cases, however, some of them have assimilated the concept of streamlined cooperation, taking place in time and space outside the organization. Indeed, during one interview, a company director was even considering making appointments during lunchtime just to "network":

> One of the problems is that everyone has their own life, but this is the same wherever you are. We don't know many people, but we're also trying to reach out… From what I see, people are quite willing, but it's more that: (1) we don't necessarily take the time to get to know others, and (2) there is little or no group spirit. I don't know if we should develop some tools and say, once a month, have lunch with someone I don't know, but who might be of some interest. We don't take the time and there

> aren't necessarily the conditions for us to take the time we want (Interview with Raphaël, accredited company director, September 2015).

However, although this rational conception of cooperation resonates with this entrepreneur, the same is not the case for most of the workers on the site. Indeed, the workers involved with laboratories and companies are aware of the programs put in place by the biocluster, but these appear as a disembodied objective, or even as a contradiction:

> For me, first of all, it's important to be clear about why we want to strengthen links. If it's just for it to be fun, we can organize soccer games, to give ourselves the illusion that there is a campus life like there might be in the US [...]. I've got the feeling that we are putting the cart before the horse. Maybe I'm wrong and Genopole has already done an in-depth study on this, but by participating in COSEB meetings, I'm trying to make an effort, to force myself to participate. But I do think it's putting the cart before the horse to set up these contacts, this activity, without asking: Why? What are we expecting from it? What is the objective? Will something come out of simply putting people in contact? I'm not saying they shouldn't do it, but they have to ask the question why [...]. If you ask companies "what are you interested in at Genopole?", they will all be politically correct and answer things like "yes, it's the dynamics". But, in fact, isn't it because we gave them some of their money? I mean money, rather than support, infrastructure and mutualization... (Interview with Patrick, research institute director, September 2015).

The above extract concludes with the priority reasons for locating businesses into the cluster, which remain support and shared facilities. Hence, the question of why people should be brought together is of primary importance to this researcher and institute director. Moreover, his phrase "putting the cart before the horse" is indicative of the nonsense seen in literature on the dynamics of geographic concentration. Indeed, in the case of clusters, the concentration of biotechnology activities is seen as a means of creating cohesion. However, this logic runs counter to traditional location strategies, both private and public, which are based on the opposite reasoning, where cohesion is a prerequisite for concentration (Lamy and Le Roux 2017, p. 97).

As we see, the system is far from being as harmonious and coherent as the cluster's communication claims. Depending on individuals' social position or sense of belonging, the relational dynamics desired by Genopole are not received and assimilated in the same way. These disparities, and even refusals to cooperate on the part of employees, both within the businesses and laboratories on the site, compromise the implementation of new organizational systems. The expression "putting the cart before the horse", used by the institute director, illustrates the implementation of an asynchronous organizational model. According to Linhart, this model consists of "changing the employees before changing the work" (Linhart 2010, p. 79). Employees are encouraged to change their behaviors, even though these new behaviors are not appropriate. It is a question, again according to Linhart, of "replacing employees, professionals and bearers of values and potentially conflicting interests with humans who are supposed to share the same issues and are considered as fellow human beings" (Linhart 2015, p. 48). In addition to maladjustment, the demand for like-minded individuals, capable of cooperating, may in fact lead social groups, with conflicting interests, to confront each other. Let us analyze a little more closely what these contradictions are about.

6.2.3. *Loyalty and performance objectives towards the employer*

Whether in laboratories or in companies, cooperation with external organizations is not a priority requirement for employers (see section 6.1). It may not even be required internally as research teams compete for funding (see Chapter 5). Cluster workers thus find themselves caught in a paradox specific to the hypermodern individual, who is both individualistic and socialized. First encountered in multinationals, this individual now finds himself at the center of contradictory demands across the working world: "on the one hand, responsible and creative, he must also fit into models (being a good learner, a graduate, feeling good, etc.), constraints (competitions, selection, hiring, etc.) and very strict standards" (de Gaulejac and Hanique 2015, p. 12). The paradox is mainly to be found in the institutional demand for a team mentality, in the company or the laboratory, while within them work is evaluated by individual performance, whether it be academic standards (publications, patents, etc.) or merit-based advancement (individual interviews, personal development, etc.).

The work organizations observed in the context of the survey all advocate values of competition, performance and personal achievement, while the cluster expects better interorganizational cohesion. The management of the biocluster is astonished by the disappearance of worker collectives, while modern management is constantly promoting individualization and rivalry. From this, we observe a dual paradoxical conflict between the contradiction between individualization and cooperation, which seems to go hand in hand with a competition paradox (encouraged by the accredited organizations), and cooperation (desired by the public intermediation structure). Indeed, Genopole reinforces this paradoxical process by demanding cohesion and cooperation from work organizations that celebrate competition and individual performance, as Castel, Hénaut and Marchal have pointed out:

> In many sectors, actors face potentially contradictory demands, with, on the one hand, incentives to perform better economically or scientifically ("be the best!"), and, on the other, demands to be more collaborative ("cooperate, share information!") (Castel *et al.* 2016, p. 10).

As we have seen, however, cooperation and information sharing are not promoted within laboratories and companies. Although the logic of cooperation was dominant before the creation of Genopole within the international "Human Genome" project, the logic of competition is now the order of the day:

> Fifteen years ago, the Human Genome Project was an international collaboration, the sequencing technologies were in such a state that everyone had to work together to produce a result, all the data was shared, everyone was assigned a role, there was true international cooperation. Now, the evolution of sequencing techniques means that we have moved to an international competition, where everyone has the capacity to generate massive data, increasingly private databases that are not shared with others. There is still shared data, but we have changed the model to a competition for access to data with large private players like Google, which is getting into the game to

> control the data and information (Interview with Patrick, research institute director, September 2015).

We can see it here: data sharing is no longer appropriate today, especially in a context of private owners such as Google, quoted in the above extract. Cluster organizations are faced with industrial giants that accumulate, protect and privatize their data. In this context, as we have seen, whether they work in laboratories or in companies, individuals in the cluster are subject to confidentiality requirements: disclosure forms, Confidential Document Agreements (CDA) for laboratories and confidentiality clauses in company contracts.

Moreover, when accredited organizations interact with other actors, they are not necessarily located nearby. An organization cooperates all the more with actors located outside the cluster when it is specialized in a particular field and therefore needs to look "elsewhere" for knowledge, skills or resources (Hussler and Ronde 2005). The following comments illustrate this:

> At the biocluster level, I have never set up any collaboration agreements, except for small contracts to use instruments in labs or in other biotechs. On the other hand, I have easily set up 30 to 40 in Europe, the United States, Canada, Brazil, Indonesia, China, Holland, Italy… […] The activities of Genopole's labs and biotechs were not useful to my development program and, as a result, I did not set up any. If there had been anyone of interest to me at Genopole, I would obviously have contacted them […]. I never made use of this proximity, because the activities of either biotechs or academics were either not useful or I had not identified them… I don't know, because at some point, when we have a need, we identify the best, based on their patents or their publications, and those who are not identifiable, well, we squeeze them out, even if they are three blocks from us (Interview with Alexandre, director of a company accredited by the biocluster, September 2015).

This business leader sheds light on the priority given to collaboration with the best in a field, based on its patents or publications, rather than based on proximity. The performance of the organization takes precedence over local cooperation. Thus, the central role of cooperation in production models, such as that of the cluster, appears to be difficult to reconcile with the individualism and competition advocated in the accredited organizations. On one side, employees are encouraged to take initiatives, to leave their organization to exchange information. On the other side, they are subject to strict standards, in terms of schedules and confidentiality, and to processes of accountability and performance with regard to their respective organizations. Finally, the legal rules of capitalism, such as private property, confidentiality and competition, and the organizational rules linked to hierarchy or modern management produce individual logics of subordination to the organization, which contradict the aims of the cluster, who would like social production to be more shared. However, cooperation is ultimately prevented by the dominant capitalist logic in the accredited organizations.

6.3. Avoidance strategies

While the raison d'être of clusters is their function of geographically grouping and connecting people, in reality their programs produce avoidance of cooperation.

6.3.1. *Avoiding scientific and technological issues*

The contradictory demands to withdraw and cooperate places individuals in delicate positions where they have to be careful about what they say and judge how much they can talk with their interlocutor:

> Company A doesn't have that confidentiality issue for example, they're into commercialization. They are no longer in the development phase of new technologies, so they are freer, they can talk more. On the other hand, Company B is more closed, we don't talk about professional things. With a colleague from the same company, we can talk about all this, because he is subject to the same secrecy as we are, and that feels good (Interview with Rizlaine, biomechanics engineer in a company accredited by the biocluster, December 2014).

In this extract, we can see the discomfort of concealing company-related information and the relief of being able to confide in a colleague. Above all, the secrecy places them in an indelicate position. Employees who are no longer bound by this confidentiality are considered "freer" insofar as they can speak. Thus, the construction of a mobilizing collective and a community of values, which is based on a juxtaposition of organizations and individuals maintained in a spirit of competition (de Gaulejac and Hanique 2015, p. 180), can have psychological consequences for already highly pressured individuals, as economic literature has also identified:

> The obligation to cooperate without the subjective conditions of this cooperation, in other words, adherence to rules and values shared in a community mode, cannot fail to have significant long-term effects. The rise in stress in the workplace, at the same time as collective reference points are disappearing and threats to employment are increasing, is weighing on the psyche of employees. The ersatz collectives represented by participation programs and company projects do not succeed in filling this void (Coutrot 1998, pp. 245–246).

Placing individuals under a constraint to cooperate, without their having the means to do so, can lead to stress and even to behaviors that are antinomic and counterproductive to the cluster's objective. During the survey, particularly during the "Site Life" and "Animation" projects, phenomena of avoidance of scientific and industrial issues were repeatedly observed. The program, one of whose objectives was to create working communities within the cluster, has often failed for this reason. We have seen this with the PhD students club, where students were forced not to reveal their thesis results by the policy of their host laboratory, which consequently fell back on organizing "professional life after thesis" meetings. The same was true of a meeting on the formation of an entrepreneurs club. A manager who had just started his business expressed this reluctance:

> I'm a bit reserved about tackling technological issues because it will be complicated. It will, out of necessity, limit the discussion. For reasons of confidentiality, I wouldn't feel free to speak... On the other hand, on questions of management, of attracting investors, that's common... But on questions of research, that's different (Extract from a meeting where the agenda was the creation of an entrepreneurs club, April 20, 2017).

In the above extract, we again note the feeling of not being free to talk on science and technology issues and the preference to avoid such topics. The 11 business leaders present at this meeting were unanimous on this point. On the contrary, they did not agree on the content of the meetings. Some wished, in a very pragmatic spirit, to address issues specific to business leaders, such as management, participation in international trade shows and finding investors, while others preferred to focus on convivial meetings (restaurants, aperitifs on a barge) that would generate confidence:

> I defend the idea that the priority is for us to create a link between us, so that we know who to turn to if we need information… For me, trust is important, so there is a balance to be found between productivity and conviviality in this group. But the priority remains conviviality, not in the social sense, but in the sense of creating a link (Extract from a meeting where the agenda was the creation of an entrepreneur's club, April 20, 2017).

The productivity aspect of the meetings is not hidden by this entrepreneur, but the creation of a bond of trust between them comes to the fore. For Simmel, social life would not be possible without "affective elements", particularly benevolence and trust (Watier 2009). This is even considered by some as a prerequisite for economic exchange (Dasgupta 2011). The extract above suggests that it would be absent from exchanges between entrepreneurs and that it would be appropriate to "create a link". Nevertheless, it remains a protean notion that does not function in a binary way ("I do or don't trust"). We can count on the benevolence of a person, without necessarily trusting them to solve a given problem (Grossetti 2009). Indeed, during one interview, a teacher–researcher explained the difficulty of talking with a colleague from another laboratory or company, even to seek help, insofar as:

> Even if I know them well, it's tricky… If you start explaining your problems, you start talking about your methods, it quickly becomes complicated… (Interview with Gerard, accredited laboratory teacher–researcher, May 2014).

As we can see, even though the interknowledge exists, the climate of trust is lacking. This context lead the 11 participants in the entrepreneurs club meeting to unanimously agree that scientific and technological subjects should be avoided, despite the intention to create forms of mutual aid between leaders. Whether between PhD students or between entrepreneurs, to take only these two examples, resistance was very present. Everything happened as if the core production activities of companies and laboratories could not be discussed, as if it were untouchable. This phenomenon was noted at the very beginning of the survey, through the consultation carried out for Genopole. One question asked about the types of activities biocluster members would like to have access to in the multi-purpose hall, which has since become L'Escale. Only 12% of the 534 people questioned for this consultation wanted to practice any scientific type of activity there. As we can see, the convivial dimension seems to be the part of the program that works best. Indeed, in addition to access to attractive leisure services, the opening of a sports and cultural activity hall seems to be one of the rare spaces where "business" or "science" is not discussed, except by chance in the course of a conversation. Everything happens as if sociability becomes possible, as soon as scientific and/or industrial considerations are avoided.

6.3.2. *Cluster administrators: between belief and lucidity*

The resistances observed above are considered by a section of French economic literature to be the unfulfilled promises of cluster policies, notably their low profitability, which does not seem to amortize the costs of the program, and the lack of correlation between spatial concentration and industrial and scientific performance (Duranton *et al.* 2010). At an individual level, frustration and guilt can form among workers who fail to participate, to cooperate with the rest of the cluster, as they strive to perform their jobs (de Gaulejac and Hanique 2015, p. 28). This guilt-tripping is all the stronger with the administrators who set up these programs and find themselves caught in the crossfire, from, on the one hand, management that expects them to produce synergies and, on the other hand, a lack of participation by individuals evolving in the accredited organizations. Indeed, faced with the failure of their mission, these workers must assume the disappointment of their management:

> I am a little disappointed […] after the excitement where there were quite a few things put in place, we see that the audiences are getting smaller and smaller. I find that a problem… When we did the first "gene cafés" in the 2000s, it was interesting. There were 250 people, but then 100 and we ended up with 10

> or 12. Was it because of the speakers? Yet we were paying attention. I never understood… I think there is an element of idea fatigue. But the very idea of animation and Site Life is to be able to renew activity, even if it means going back to the original ideas. It seems to me that there is a need for renewal that is not done enough (Interview with biocluster management, April 2017).

In this interview extract, we note the process of renewal of the programs, since the 2000s, which initially aroused enthusiasm and then suffered from a lack of participation, creating disappointment among the administrators. This difficulty for the "Administrative" and "Site Life" teams in encouraging participation among cluster members was observed at close quarters in the context of the CIFRE. The findings of the survey on the structural dysfunction of the facility over the past 15 years were also shared with the teams. They were relieved to understand that the problem did not come from their work per se. Like Dujarier's "planners", they were not fooled and they too noted the discrepancy between the promises of the plan they had drawn up and the reality on the ground. In spite of the disappointment observed among the managers, we also observe phenomena of lucidity with regard to the programs put in place:

> The first level is: do I believe in partnership policy as a component of innovation development? The answer is "No". It allows people to get to know each other and develop joint research […] And, in fact, when you look at the economic and industrial results of all this, there is not much to show for the money we put in. On the lab side, the lab says "what I am interested in is having a nice publication", the industrialist says "it is the patent that I am interested in". But he doesn't always take the patent, because it's not necessarily part of the company strategy. When you launch a project, it has to be a project you care about, but, after three years, your strategy may have changed completely […]. In this way, the partnership policy does not seem convincing to me. Secondly, money: if you want to set up calls for projects, you need significant resources and here, again, you need to find funding. The only thing I believe in is Grant Services, not as a funding organization, but as an organization that allows you to find outside funding, although

> not us directly… (Interview with biocluster management, April 2017).

This extract highlights several aspects. First, the administrators of the cluster believe that the partnership policy has limited effects in terms of economic spin-offs, insofar as they note the different interests of laboratories and businesses. Moreover, the above statement underlines the awareness that, without a call for projects that mobilizes actors in a funding logic, it is very difficult to set up partnerships. Thus, we observe a form of lucidity on the limited effects of the cooperation mechanism. However, this lucidity does not go hand in hand with a belief in the dominant discourse. Indeed, the field study allowed us to observe that all the staff of the intermediation structure oscillated between belief in the cluster effect and lucidity regarding the difficulties they encounter on a daily basis, between, on the one hand, the sugar-coating of communication (see Chapter 3) and, on the other hand, the awareness of the program's limitations. In so doing, they are constrained, and also want to believe in it. This belief, linked to the cluster's imaginary, is indispensable to the cluster's administrators in order to find meaning to their activity.

Although the networking program has been formalized since the 2010s, Genopole has been trying to increase synergies since its inception, as we have seen throughout this book. The renewal of the same process, albeit applied to different audiences (bringing together managers, immunology specialists and PhD students), is part of an unconscious practice of masking the obstacles to cooperation. Indeed, in order to fit in with the dominant discourse, that of the knowledge-based economy and harmonious interactions between science and industry, administrators must produce their own *ad hoc* rhetoric and put in place mechanisms that respond to these representations. Thus, the difficulties in putting them in place are identified, but mitigated by belief, and the need to believe, in this imaginary of clustering. This belief allows the program to constantly renew, and also to provide these administrators with a meaning to their work.

Conclusion

Recent clustering policies, marked in France since 2004 by the competitiveness clusters program, are presented as unprecedented, bringing about change in a sclerotic world where science and industry are insufficiently linked. As we reach the end of this book, we understand that clustering policies succeed in changing space by geographically grouping research and production activities. From there, it would seem that geographical proximity can be rationalized by means of accreditation strategies and service offers provided by public authorities. However, it seems less obvious to initiate cooperation between these accredited members. On the contrary, clustering generates multiple contradictions. Although it is thought to bring together groups and individuals with similar activities, through disciplinary qualifications, such as in biotechnology, or identities, such as the creative class, clustering nevertheless creates stresses between social fields with distinct logics and boundaries. Although public innovation policies attempt to blur these boundaries, particularly within the scientific field where researchers are encouraged to valorize their research through industrial collaboration or the creation of businesses, resistance remains.

Beyond the fact that it has a limited impact in terms of cooperation, the clustering process is paradoxical in several contexts, insofar as geographic clustering brings together distinct social groups. The constraint to cooperate places them in a contradictory position where they must, on the one hand, respond to legal and managerial norms specific to their respective organizations and, on the other hand, adopt a behavior of autonomous, flexible working, with freedom to participate in any event. When put to the test, the actors show resistance through their lack of participation and forbidden cooperative efforts. Finally, this work shows that the cooperation promoted by cluster policies is hardly compatible with the phenomena of individualization and competition observed in the various practices of scientific work. Individuals even go so far as to redefine the program by putting aside the main expectations of the cluster in terms of cooperation: the scientific and technological dimensions. They thus avoid cooperation.

In this context, clusters convey an innovation imaginary on two levels: outside and inside the cluster. The real and figurative real estate expansion, as shown in the aerial maps of the studied cluster, is a good example. Expanding geographically is visible proof of the territorialization of the cluster for the collectives and individuals located there, and allows it to attract companies and laboratories outside the cluster, as well as to increase or maintain funding. While the cooperative dynamics of the cluster remain patchy, the rhetoric is strengthened over time. Its operability is such that it manages to mask the renewed character of clusters. Indeed, on a national scale, it reproduces similar policies to those conducted in the past with the nuclear and aeronautics industries (Lamy and Le Roux 2017, p. 89). At the local level, the process of linking is repeated over and over again, without showing convincing results at each attempt. The simple decreeing of clustering should be enough to produce its effects, like an irrefutable scientific law. Questioning appears almost impossible for institutions and individuals outside the cluster. On the contrary, inside the cluster, we observe phenomena of lucidity on the part of the administrators who organize the connections, but this clear-sightedness is accompanied by a strong belief in clustering. They remain convinced that, in the long run, by dint of perseverance and reiteration, interactions between individuals within the cluster will no longer need to be encouraged but will happen by themselves. The power of the imaginary thus takes precedence over reality. The cluster is not only naturalized, but it also acts as a self-fulfilling prophecy.

By making efforts to persuade and rationalize, clusters succeed in making the development of already proven industrial models new and essential for economic progress. Moreover, clusters manage to rationalize social systems that were originally spontaneous, as in the Italian district during the days of the technopoles in France. As a result of this rational logic, entire campuses are emerging, disconnected from the territories in which they are located. This rationalization is made possible by the accreditation, which acts as additional capital. This accredited capital provides companies and laboratories with multiple advantages and offers the possibility of prescribing the geographical grouping of different organizations.

While this grouping has a limited impact on local cooperation, it does have an impact on the external collaboration of cluster members. The old hierarchies of scale no longer apply and a very small business can interact directly with other very small businesses around the globe (Sassen 2009, p. 237). Each business has its own regional, national, European and so on, geography; however, globalization is embodied in local physical locations (Sassen 2009, p. 237). Clusters are part of this and enable national and international exchanges to be based on local materiality, especially public intermediation structures that help the development of companies and laboratories, but that also help provide access to state-of-the-art equipment, which is indispensable for innovation activities. Infrastructure, equipment, institutions and the qualified labor market are anchored locally, but at the service of each accredited member, to develop its collaborations in a globalized economy. Indeed, the continuity of flows paradoxically requires fixed spatial configurations (Harvey 2008).

In the end, there can be no control of local cooperation by the cluster since it takes place elsewhere. Also, these external exchanges are reinforced by the various tools provided by clusters (fundraising, accreditation, territorial marketing, equipment, etc.). Accredited companies and laboratories benefit from this to collaborate within the global biotechnology market and, to a lesser extent, to cooperate locally. The latest example of the impact of accreditations for research and innovation activities is the 14th place classification of the Paris-Saclay University in the Shanghai 2020 rankings. This is the first time in 17 years that a French university has made it into the top 15. This position is mainly due to the grouping of several universities and research organizations in the south of Paris region, which represent 15% of French research, but which alone have never been ranked. The "Paris-Saclay" accreditation does not act as a vector of local dynamism, but essentially as a guarantee of the quality and performance of its institutions at the global level. In a globalized and competitive world, geographical grouping is mainly mobilized to give rise to economic and scientific giants. Their success criteria are based on the revenues of companies and the number of scientific publications (or the number of Nobel Prize winners and Fields Medalists, in the case of the Shanghai ranking), rather than on the benefits of a possible "cluster effect".

References

Abdelnour, S. (2014). L'auto-entrepreneuriat : une gestion individuelle du sous-emploi. *La nouvelle revue du travail*, 5, 2 November.

Achermann, G. (2014). Le modèle de la "triple hélice" appliqué au territoire : problématique, limites et intégration de nouvelles variables explicatives. Working Papers, 285, Laboratoire de recherche sur l'industrie et l'innovation, ULCO/Research Unit on Industry and Innovation.

Aggeri, F., Genet, C., Lanciano-Morandat, C., Mangematin, V. (2007a). Modèle d'activité dans l'instrumentation en biotechnologies. Construire l'offre ou répondre à la demande ? *Revue d'économie industrielle*, 120, 41–60, 15 December.

Aggeri, F., Le Masson, P., Branciard, A., Paradeise, C., Peerbaye, A. (2007b). Les plates-formes technologiques dans les sciences de la vie. *Revue d'économie industrielle*, 120, 21–40, 15 December.

Ahrne, G. and Brunsson, N. (2008). *Meta-organizations*. Edward Elgar Pub, Cheltenham.

Aït Ali, N. and Rouch, J.P. (2013). Le "je suis débordé" de l'enseignant-chercheur. Petite mécanique des pressions et ajustements temporels. Temporalités. *Revue de sciences sociales et humaines*, 18, 19 December.

Akrich, M. (1993). Les formes de la médiation technique. *Réseaux*, 60, 87–98.

Akrich, M., Callon, M., Latour, B. (1988). À quoi tient le succès des innovations ? 1: L'art de l'intéressement ; 2: Le choix des porte-parole. Gérer et comprendre. *Les Annales des mines*, 4–17.

Alain, E. and Schaffar, L. (1995). Organisation de la recherche et conformisme scientifique. *Population*, 50(2), 509–512.

Allègre, C. (1993). *L'Âge des savoirs : pour une renaissance de l'Université*. Gallimard, Paris.

Alter, N. (2000). *L'innovation ordinaire*. Presses universitaires de France, Paris.

Alter, N. (2002). Théorie du don et sociologie du monde du travail. *Revue du MAUSS*, 20, 263–285.

Amsellem, N., Chevallier, M., Durand, C., Heil, C., Lacroix, G., Piazza, B. (2002). Regards sociologiques sur les biotechnologies. Université d'Évry Val d'Essonne, Évry.

Appold, S. (2005). Location patterns of US industrial research: Mimetic isomorphism and the emergence of geographic charisma. *Regional Studies*, 39(1), 17–39.

Aubert, N. (1998). Relations de pouvoir et leadership. In *Traité de sociologie du travail*, De Coster, M. and Pichault, F. (eds). De Boeck, Brussels.

Aubert, N. (2006). L'individu hypermoderne et ses pathologies. *L'information psychiatrique*, 82(7), 605–610.

Aubert, N. and de Gaulejac, V. (2007). *Le Coût de l'excellence : new edition*. Le Seuil, Paris.

Audetat, M. (2015a). *Sciences et technologies émergentes : pourquoi tant de promesses ?* Hermann, Paris.

Audetat, M. (2015b). Dessine moi un futur ! Enjeux et ambiguïtés du régime des promesses scientifiques. *Sciences et technologies émergentes : pourquoi tant de promesses ?* Hermann, Paris.

Audretsch, D. and Stephan, P. (1996). Company-scientist locational links: The case of biotechnology. *American Economic Review*, 86(3), 641–652.

Azam, G. (2008). Les droits de propriété sur le vivant. *Développement durable et territoires. Économie, géographie, politique, droit, sociologie*, dossier 10, 7 March.

Azoulay, P., Ding Waverly, W., Stuart, T. (2009). *The Impact of Academic Patenting on the Rate, Quality and Direction of (Public) Research Output*. Social Science Research Network, Rochester.

Bachelard, G. (1993). *L'eau et les rêves : essai sur l'imagination de la matière*. Le Livre de Poche, Paris.

Badillo, P.Y. (2013). Les théories de l'innovation revisitées : une lecture communicationnelle et interdisciplinaire de l'innovation ? Du modèle "Émetteur" au modèle "communicationnel". *Les Enjeux de l'information et de la communication*, 14/1(1), 19–34.

Bagla, L. (2011). VI. Limites du "management culturel et participatif". *Sociologie des organisations*. La Découverte, Paris.

Bagnasco, A. (1977). *Tre Italie: La problematica territoriale dello sviluppo italiano.* Il Mulino, Bologna.

Bagnasco, A. and Triglia, C. (1988). *La construction sociale du marché. Le défi de la Troisième Italie*. ENS Cachan, Gif-sur-Yvette.

Barbusse, B. (2008). Sport et entreprise : des logiques convergentes ? *L'Année sociologique*, 52(2), 391–415, 1 February.

Barral, C. and Paterson, F. (1994). L'Association française contre les myopathies : trajectoire d'une association d'usagers et construction associative d'une maladie. *Sciences sociales et santé*, 12(2), 79–111.

Barrier, J. (2014). Partenaires particuliers : financements sur projet et travail relationnel dans les réseaux de collaboration science-industrie. *Genèses*, 94, 55–80, 10 June.

Barrier, J. and Vézian, A. (2015). Concurrence, rhétorique et mécano institutionnel : les usages stratégiques des dispositifs d'organisation de la recherche. In *Faire la concurrence. Retour sur un phénomène social et économique*, Castel, P., Hénaut, L., Marchal, E. (eds). Presses des Mines, Paris.

Barthe, J.-F., Chauvac, N., Grossetti, M. (2016). Entrepreneurs de circonstance : une enquête sociologique sur les fondateurs de start-up en France. *Revue de l'Entrepreneuriat*, 15(3), 163–180, 2 December.

Bathelt, H., Malmberg, A., Maskell, P. (2004). Clusters and knowledge: Local buzz, global pipelines and the process of knowledge creation. *Progress in Human Geography*, 28(1), 3156.

Becattini, G. (1989). Riflessioni sul distretto industriale marshalliano come conceto socio-economico. *Stato e Mercato*, 111–128, April.

Becker, H.S. (2002). *Les ficelles du métier*. La Découverte, Paris.

Béhar, D., Estèbe, P., Rio, N. (2013). La mixité économique comme volonté et comme représentation. Des villes nouvelles aux clusters en Île-de-France. 1963-2013. Report, PUCA.

Bellet, M., Colletis, G., Lung, Y. (eds) (1993). Introduction au numéro spécial "Économie de proximités". *Revue* d'Économie Régionale et Urbaine, 3, 357–361.

Bellet, M., Kirat, T., Largeron, C. (eds) (1998). *Approches multiformes de la proximité*. Hermès, Paris.

Benko, G. (2007). Territoires et sciences sociales. In *Régimes territoriaux et développement économique*, Itçaina, X., Palard, J., Ségas, S. (eds). Presses universitaires de Rennes, Rennes.

Benko, G. and Lipietz, A. (1992). *Les Régions qui gagnent. Districts et réseaux : les nouveaux paradigmes de la géographie économique*. Presses universitaires de France, Paris.

Benko, G., Dunford, M., Lipietz, A. (1996). Les districts industriels revisités. In *Dynamiques territoriales et mutations économiques*, Pecqueur, B. (ed.). L'Harmattan, Paris.

Benner, M. and Sandström, U. (2000). Institutionalizing the triple helix: Research funding and norms in the academic system. *Research Policy*, 29(2), 291–301.

Bensaude-Vincent, B. (2015). Promesse et régime d'historicité en technosciences. In *L'innovation en eaux troubles. Sciences, techniques, idéologies*, Sainsaulieu, I. and Saint-Martin, A. (eds). Éditions du Croquant, Vulaines-sur-Seine.

Bergeron, H., Castel, P., Dubuisson-Quellier, S. (2014). Gouverner par les labels. *Gouvernement et action publique*, 3, 7–31, 20 October.

Bernardy, M. and Boisgontier, P. (1988). *Grains de technopoles, micro-entreprises grenobloises et nouveaux espaces productifs*. Presses universitaires de Grenoble, Grenoble.

Bernatchez, J. (2010). La valorisation commerciale de la recherche universitaire. Principes, modalités et enjeux d'éthique publique. *Éthique publique. Revue internationale d'éthique sociétale et gouvernementale*, 12(1), 55–78, 28 December.

Bernela, B. and Levy, R. (2017). Collaboration networks within a French cluster: Do partners really interact with each other? *Papers in Regional Science*, 96(1), 115–138.

Bès, M.P. and Grossetti, M. (2001). Encastrements et découplages dans les relations science-industrie. *Revue française de sociologie*, 42(2), 327–355.

Bès, M.P. and Grossetti, M. (2002). Proximité spatiale et relations science-industrie : savoirs tacites ou encastrement (Polanyi ou Polanyi). *Revue d'économie régionale et urbaine*, 5, 777–778, December.

Bès, M.P. and Grossetti, M. (2003). Dynamiques des réseaux et des cercles. Encastrements et découplages. *Revue d'économie industrielle*, 103(1), 43–58.

Bessin, M., Bidart, C., Grossetti, M. (2010). *Introduction générale. L'enquête sur les bifurcations : une présentation*. La Découverte, Paris.

Bidart, C. (1988). Sociabilités : quelques variables. *Revue française de sociologie*, 29(4), 621–648.

Bidart, C. (1991). L'amitié, les amis, leur histoire. Représentations et récits. *Sociétés contemporaines*, 5, 21–42.

Bidart, C. (2012). Réseaux personnels et processus de socialisation. *Idées économiques et sociales*, 169, 8–15, 25 December.

Bidart, C., Degenne, A., Grossetti, M. (2011). *La vie en réseau. Dynamique des relations sociales*. Presses universitaires de France, Paris.

Blanc, C. (2004). Pour un écosystème de la croissance. Report, La Documentation française, Paris.

Blanchot, F. and Fort, F. (2007). Coopétition et alliances en R&D. *Revue française de gestion*, 176, 163–181, 22 November.

Boltanski, L. and Chiapello, È. (1999). *Le nouvel esprit du capitalisme*. Gallimard, Paris.

Bonte, P. and Izard, M. (2010). *Dictionnaire de l'ethnologie et de l'anthropologie*, 4th edition. Presses Universitaires de France, Paris.

Borup, M., Brown N., Konrad, K., Van Lente, H. (2006). The sociology of expectations in science and technology. *Technology Analysis & Strategic Management*, 18(3–4), 285–298.

Borzèix, A. and Linhart, D. (1988). La participation : un clair-obscur. *Sociologie du travail*, 30(1), 37–53.

Boschma, R. (2005). Proximity and innovation: A critical assessment. *Regional Studies*, 39, 61–74, 1 February.

Boschma, R. and Frenken, K. (2009). *The Spatial Evolution of Innovation Networks: A Proximity Perspective*. Group Economic Geography, Utrecht.

Bouba-Olga, O. and Grossetti, M. (2010). Socio-économie de proximité. *Revue d'économie régionale et urbaine*, 3, 311–328, 1 January.

Bouchard, G. (2013). Pour une nouvelle sociologie des mythes sociaux. *Revue européenne des sciences sociales*, 51–1, 95–120, 1 June.

Bouchez, J.P. (2004). *Les nouveaux travailleurs du savoir*, 1st edition. Éditions d'Organisation, Paris.

Bouchez, J.P. (2016). *L'entreprise à l'ère du digital. Les nouvelles pratiques collaboratives*. De Boek Université, Louvain-la-Neuve.

Bouinot, J. (2007). Les pôles de compétitivité : le recours au modèle des clusters ? *Cybergeo: European Journal of Geography*, 9 March.

Bourdieu, P. (1976). Le champ scientifique. *Actes de la recherche en sciences sociales*, 2(2), 88–104.

Bourdieu, P. (1980). Le capital social. Notes provisoires. *Actes de la recherche en sciences sociales*, 31, 2–3, January.

Bourdieu, P. (1984). La délégation et le fétichisme politique. *Actes de la recherche en sciences sociales*, 52(1), 49–55.

Bourdieu, P. (1989). *La noblesse d'État. Grandes écoles et esprit de corps.* Les Éditions de Minuit, Paris.

Bourdieu, P. (2013). Séminaires sur le concept de champ, 1972–1975. *Actes de la recherche en sciences sociales*, 200, 4–37.

Bourque, R. and Thuderoz, C. (2002). *Sociologie de la négociation*. La Découverte, Paris.

Boyer, R. and Saillard, Y. (2002). *Théorie de la régulation, l'état des savoirs.* La Découverte, Paris.

Branciard, A. (1999a). Espace d'innovation dans la biologie et recomposition d'espaces productifs. Analyse des processus institutionnels et politiques en œuvre. Report, CNRS, Paris.

Branciard, A. (1999b). L'articulation science/innovation en France : dix ans d'action publique pour le développement de la génomique et des biotechnologies. *Les Cahiers de l'innovation*, 05/2000.

Branciard, A. (2002). Politiques publiques et espace d'innovation dans la biologie. Étude de dispositifs d'intégration science/industrie et de création d'entreprises : le cas de la Génopole d'Évry. Report, HAL.

Branciard, A. (2004). Le développement économique lié aux potentiels scientifique et technologique en génomique : action publique nationale et dynamique régionale en PACA. Le cas de Marseille Nice Génopole. Report, HAL.

Branciard, A. (2009). L'invention de Genopole®, emblème et acteur d'une nouvelle politique de recherche et d'innovation en France. *Médecine/sciences*, 25(2), 9–12, May.

Branciard, A. and Crespy, C. (2007). Des dispositifs d'action publique territorialisés vecteurs de reconfiguration des politiques scientifique et technologique en France. *Deuxième congrès de l'Association française de sociologie Réseau 29.*

Branscomb, L. and Keller, J. (1998). *Investing in Innovation: Creating a Research and Innovation Policy that Works*. MIT Press, Cambridge.

Brenner, N. (2004). *New State Spaces: Urban Governance and the Rescaling of Statehood*. OUP Oxford, Oxford/New York.

Brown, J.S. and Cook, S.D.N. (1999). Bridging epistemologies: The generative dance between organizational knowledge and organizational knowing. *Organization Science*, 10, 381–400.

Brown, N. and Michael, M. (2003). A sociology of expectations: Retrospecting prospects and prospecting retrospects. *Technology Analysis and Strategic Management*, 15(1), 4–18.

Brucy, G. (2012). Former pour rendre “employable” ? Genèse d’une politique. In *La professionnalisation de l’enseignement supérieur. De la volonté politique aux formes concrètes*, Quenson, E. and Coursaget, S. (eds). Octarès, Toulouse.

Brunet, P. (2011). Procès de production d’une recherche orientée entre chercheurs, malades et entrepreneurs : le cas d’une “biotech à but non lucratif”. *Revue Interventions économiques*, 43, 1 May.

Brunet, P. (2012). Temporalités dans la recherche biomédicale : la science au travail saisie par le temps. *La nouvelle revue du travail*, 1, 10 December.

Brunet, P. and Dubois, M. (2012). Cellules souches et technoscience : sociologie de l’émergence et de la régulation d’un domaine de recherche biomédicale en France. *Revue française de sociologie*, 53(3), 391–428.

Brunet, R., Grasland, L., Garnier, J.P., Ferras, R., Voile, J.P. (1988). *Montpellier Europole*. GIP RECLUS/Maison de la géographie, Montpellier.

Bruno, I. (2008). *À vos marques, prêts... cherchez ! La stratégie européenne de Lisbonne, vers un marché de la recherche*. Éditions du Croquant, Paris.

Bruno, I. (2009). Le temps des “chercheurs-entrepreneurs” : sens et pouvoir du benchmarking dans l’“espace européen de la connaissance”. *Quaderni*, 69, 93–104.

Bruno, I. and Didier, E. (2013). *Benchmarking*. Zones, Paris.

Brusco, S. (1982), The Emilian model: Productive decentralisation and social integration. *Cambridge Journal of Economics*, 6, 167–184.

Brusco, S. (1986). Small firms and industrial districts: The experience of Italy. In *New Firms and Regional Development in Europe*, Keeble, D., Wever, E. (eds). Routledge, Beckenham.

Brusco, S. (1990). The idea of the industrial district: Its genesis. In *Industrial Districts and Inter-firm Cooperation in Italy*, Pyke, F., Becattini, G., Sengenberger, W. (eds). International Institute for Labour Studies, Geneva.

Buquet, R., Luca, N., Bouilloud, J.P. (2017). Malaise dans les start-up. Entre désir héroïque et anxiété créatrice. *Nouvelle revue de psychosociologie*, 24, 93–111, 28 November.

Burt, R.S. (1992). *Structural Holes: The Social Structure of Competition*, 1st paperback edition. Harvard University Press, Cambridge.

Burt, R.S. (2007). *Brokerage and Closure: An Introduction to Social Capital*. Oxford University Press, Oxford.

Callon, M. (1986). Éléments pour une sociologie de la traduction. La domestication des coquilles Saint-Jacques et des marins pêcheurs dans la baie de Saint-Brieuc. *L'Année sociologique*, 36, 169–208.

Callon, M. (1998). An essay on framing and overflowing: Economic externalities revisited by sociology. *The Sociological Review*, 46, 244–269.

Callon, M. and Ferrary, M. (2006). Les réseaux sociaux à l'aune de la théorie de l'acteur-réseau. *Sociologies pratiques*, 2(13), 37–44.

Callon, M. and Rabeharisoa, V. (1999). *Le pouvoir des malades : l'Association française contre les myopathies et la recherche*. Presses des Mines, Paris.

Callon, M. and Rabeharisoa, V. (2002). L'engagement des associations de malades dans la recherche. *Revue internationale des sciences sociales*, 171, 65–73.

Callon, M., Joly, P.-B., Rip, A. (2010). Reinventing innovation. *Governance and Innovation*, 19–32.

Caloghirou, Y., Tsakanikas, A., Vonortas, N. (2001). University-industry cooperation in the context of the European framework programmes. *The Journal of Technology Transfer*, 26, 153–161.

Camagni, R. and Maillat, D. (2006). *Milieux innovateurs. Théorie et politiques*. Economica, Paris.

Cangilhem, G. (2000). *La connaissance de la vie*. Librairie philosophique Vrin, Paris.

Capecchi, V., Costa, G., Novelli, A. (1987). Formation professionnelle et petite entreprise : le développement industriel à spécialisation flexible en Émilie-Romagne. *Formation Emploi*, 19(1), 3–18.

Carayol, N. (2003). Objectives, agreements and matching in science-industry collaborations: Reassembling the pieces of the puzzle. *Research Policy*, 32(6), 887–908.

Carnino, G. (2014). Louis Pasteur. La science pure au service de l'industrie. *Le Mouvement social*, 248, 9–26, 8 September.

Carnino, G. (2015). *L'invention de la science. La nouvelle religion de l'âge industriel*. Le Seuil, Paris.

Caroly, S. and Clot, Y. (2004). Du travail collectif au collectif de travail : développer des stratégies d'expérience. *Formation Emploi*, 88, 43–55.

Carré, D. and Levratto, N. (2011). *Les performances des territoires : les politiques locales, remèdes au déclin industriel*. Le Manuscrit, Paris.

Cassier, M. (1996). Les contrats de recherche entre l'université et l'industrie : les arrangements pour la production de biens privés, de biens collectifs et de biens publics. *Sociologie du travail*, 38(3), 377–390.

Cassier, M. (2002a). L'engagement des chercheurs vis-à-vis de l'industrie et du marché : normes et pratiques de recherches dans les biotechnologies. In *Les logiques de l'innovation. Approche pluridisciplinaire*, Alter, N. (ed.). La Découverte, Paris.

Cassier, M. (2002b). Paul Rabinow. Le déchiffrage du génome, l'aventure française. *Sciences sociales et santé*, 20(3), 125–130.

Castel, P., Hénaut, L., Marchal, E. (2016). *Faire la concurrence : retour sur un phénomène social et économique*, 1st edition. Presses des Mines, Paris.

Castells, M. (2001). *La société en réseaux*. Fayard, Paris.

Castells, M. and Hall, P. (1994). *Technopoles of the World: The Making of 21st Century Industrial Complexes*. Routledge, London.

Castoriadis, C. (1999). *L'institution imaginaire de la société*. Le Seuil, Paris.

Cederström, C. and Spicer, A. (2016). *Le syndrome du bienêtre*. L'Échappée, Paris.

de Certaines, J. (1988). *La fièvre des technopoles*. Syros Alternatives, Paris.

Chamboredon, J.C. and Lemaire, M. (1970). Proximité spatiale et distance sociale. Les grands ensembles et leur peuplement. *Revue française de sociologie*, 11(1), 3–33.

Chiapello, E. (1998). *Artistes versus managers : le management culturel*. Métailié, Paris.

Clerval, A. (2011). David Harvey et le matérialisme historico-géographique. *Espaces et sociétés*, 147, 173–185.

Cohen, E. (1992). *Le colbertisme high tech. Économie des télécom et du grand projet*. Hachette, Paris.

Cohen, D. (2006). *Trois leçons sur la société post-industrielle*. Le Seuil, Paris.

Coleman, J.S. (1994). *Foundations of Social Theory*. Harvard University Press, Harvard.

Comité de la gestion publique OCDE (1994). La gestion des performances dans l'administration : mesure des performances et gestion axée sur les résultats. Report, Organisation de coopération et de développement économique, Paris.

Commission permanente du conseil régional (2017). Contribution 2017 au Groupement d'intérêt public Genopole. Report presented by Valérie Pécresse, Commission permanente du conseil régional, January, Paris.

Conseil de l'Union européenne (2003). Investir efficacement dans l'éducation et la formation. Report, Conseil de l'Union européenne, Brussels.

Conseil de l'Union européenne (2004). Éducation et Formation 2010 : l'urgence des réformes pour réussir la stratégie de Lisbonne. Report, Conseil de l'Union européenne, Brussels.

Cooke, P. (2001). *Regional Innovation Systems, Clusters, and the Knowledge Economy*. Oxford University Press, Oxford.

Coppin, O. (2002). Le milieu innovateur : une approche par le système. *Innovations*, 16(2), 29–50.

Coriat, B. (2002). Le nouveau régime américain de la propriété intellectuelle. Contours et caractéristiques clés. *Revue d'économie industrielle*, 99(1), 17–32.

Courlet, C. (2002). Les systèmes productifs localisés. *Études et recherches sur les systèmes agraires et le développement*, 27–40.

Courlet, C. and Pecqueur, B. (1992). Les systèmes industrialisés localisés en France : un nouveau modèle de développement. In *Les régions qui gagnent*, Benko, G., Lipietz, A. (eds). PUF, Paris.

Coutrot, T. (1998). *L'entreprise néolibérale, nouvelle utopie capitaliste ? Enquête sur les modes d'organisation du travail*. La Découverte, Paris.

Crespo, M. and Dridi, H. (2007). Intensification of university–industry relationships and its impact on academic research. *Higher Education*, 54(1), 61–84.

Crespy, C. (2007). Gouvernance de la recherche et compétitivité des régions : quel rôle pour l'action publique territoriale ? *Politiques et management public*, 25(2), 23–44.

Crozier, M. (1991). *L'entreprise à l'écoute, apprendre le management postindustriel*. InterEditions, Paris.

Crozier, M. and Friedberg, E. (1977). *L'acteur et le système : les contraintes de l'action collective.* Le Seuil, Paris.

Dagnaud, M. (2016). *Le modèle californien*. Odile Jacob, Paris.

Dahan, A. and Mangematin, V. (2010). Recherche, ou temps perdu ? Vers une intégration des tâches administratives au métier d'enseignant-chercheur. *Annales des mines. Gérer et comprendre*, 102, 14–24.

Dardot, C. and Weber, L. (2002). *Le nouvel ordre éducatif mondial. OMC, Banque mondiale, OCDE, Commission européenne*. Syllepses, Paris.

Dasgupta, P. (2011). Une question de confiance : capital social et développement économique. *Revue d'économie du développement*, 18(4), 47–96, 10 February

Dasgupta, P. and David, P. (1994). Toward a new economics of science. *Research Policy*, 23(5), 487–521.

Daumas, J.C. (2007). Districts industriels : du concept à l'histoire. Les termes du débat. *Revue économique*, 58(1), 131–152.

David, P. and Foray, D. (2002). Une introduction à l'économie et à la société du savoir. *Revue internationale des sciences sociales*, 171, 13–28.

Dejours, C. (2013). *Travail vivant. Tome II : Travail et émancipation*. Payot, Paris.

Del Río Carral, M. and Fusulier, B. (2013). Jeunes chercheurs face aux exigences de disponibilité temporelle. Quelles logiques d'agencement entre la vie professionnelle et la vie privée ? *Temporalités. Revue de sciences sociales et humaines*, 18, 19 December.

Depret, M.H. and Hamdouch, A. (2009). Clusters, réseaux d'innovation et dynamiques de proximité dans les secteurs high-tech. Une revue critique de la littérature récente. *Revue d'économie industrielle*, 128, 21–52, 15 December.

Dibiaggio, L. and Ferrary, M. (2003). Communautés de pratique et réseaux sociaux dans la dynamique de fonctionnement des clusters de hautes technologies. *Revue d'économie industrielle*, 103(1), 111–130.

DiMaggio, P.J. and Powell, W.W. (1983). The iron cage revisited: Institutional isomorphism and collective rationality in organizational fields. *American Sociological Review*, 48(2), 147–160.

Dosi, G. (1998). The nature of the innovation process. In *Technical Change and Economic Theory*, Dosi, G., Nelson, R., Silverberg, G., Soete, L. (eds). Pinter Publishers, London.

Drucker, P. (2010). *L'avenir du management : réflexions pour l'action*. Pearson Village Mondial, Paris.

Dujarier, M.A. (2015). *Le management désincarné*. La Découverte, Paris.

Dupuy, C. and Burmeister, A. (eds) (2003). *Entreprises et territoires, les nouveaux enjeux de la proximité*. La Documentation Française, Paris.

Dupuy-Maury, F. (2009). Genopole® : une genèse de dix longues années. *Médecine/sciences*, 25, 5–8, 1 February.

Durand, J.P. (1991). Italie : spécialisation flexible et dépassement du fordisme. *Revue d'économie industrielle*, 58(1), 47–63.

Durand, J.P. (2012). *La chaîne invisible. Travailler aujourd'hui : flux tendu et servitude volontaire*. Le Seuil, Paris.

Durand, J.P. (2016). Richard Florida, *The Rise of the Creative Class, Revisited*, New York, Basic Books, 2012, 484 p [Online]. Available at: http://journals.openedition.org/nrt/2971.

Durand, G. and Cazenave, M. (1996). *Introduction à la mythodologie : mythes et sociétés*. Albin Michel, Paris.

Durand, J.P., Tiffon, G., Moatty, F. (2014). *L'innovation dans le travail*. Octarès, Toulouse.

Duranton, G., Martin, P., Mayer, T., Mayneris, F. (2008). *Les pôles de compétitivité. Que peut-on en attendre ?* Éditions Rue d'Ulm, Paris.

Duranton, G., Martin, P., Mayer, T., Mayneris, F. (2010). *The Economics of Clusters: Lessons from the French Experience*. OUP Oxford, Oxford.

Eckert, D., Grossetti, M., Martin-Brelot, H. (2012). La classe créative au secours des villes ? *La vie des idées*, 28 February.

Elias, N. (1991). *Qu'est-ce que la sociologie ?* Éditions de l'Aube, Paris.

Elias, N. (1993). *Engagement et distanciation : contributions à la sociologie de la connaissance*. Fayard, Paris.

Ellul, J. (2012). *Le système technicien*. Cherche Midi, Paris.

Eloire, F., Penalva Icher, É., Lazega, E. (2012). Application de l'analyse des réseaux complets à l'échelle interorganisationnelle. *Terrains et travaux*, 19, 77–98, 6 February.

Epstein, R. (2006). Gouverner à distance. *Esprit*, 11, 96–111, November.

Estades, J., Joly, P.B., Mangematin, V. (1996). Dynamique des relations industrielles dans les laboratoires d'un grand organisme public de recherche : coordination, apprentissage, réputation et confiance. *Sociologie du Travail*, 3, 391–407.

Etzkowitz, H., Asplund, P., Nordman, N. (2003). Beyond Humboldt: The entrepreneurial university the third mission and the triple helix. *VEST Journal for Science and Technology Studies*, 16(1), 21–45.

Faberon, J.Y. (1990). Technopoles et développement. *Revue française de science politique*, 40(1), 46–63.

Fallon, C. and Thoreau, F. (2016). Rejouer les partitions disciplinaires : l'interdisciplinarité comme épreuve dans le cas de la génomique. In *Le gouvernement des disciplines académiques. Acteurs, dynamiques, instruments, échelles*, Benninghoff, M. (ed.). Éditions des archives contemporaines, Paris.

Faulkner, W. and Senker, J. (1994). Making sense of diversity: Public-private sector research linkage in three technologies. *Research Policy*, 23, 673–695.

Faure, S. and de Montlibert, C. (2005). Savoirs à vendre. L'enseignement supérieur et la recherche en danger. *Lectures*, 18 September.

Favret-Saada, J. (1977). *Les mots, la mort, les sorts. La sorcellerie dans le bocage.* Gallimard, Paris.

Feldman, M.P., Johanna, F., Janet, B. (2005). Creating a cluster while building a firm: Entrepreneurs and the formation of industrial clusters. *Regional Studies*, 39(1), 129–141, February.

Felt, U. (1993). Fabricating scientific success stories. *Public Understanding of Science*, 2, 375–390, 1 October.

Ferrary, M. and Pesqueux, Y. (2004). *L'organisation en réseau, mythes et réalités*, 1st edition. Presses universitaires de France, Paris.

Fischer, C. (2011). *Still Connected: Family and Friends in America Since 1970.* Russell Sage Foundation, New York.

Flichy, P. (2001). *L'imaginaire d'Internet.* La Découverte, Paris.

Flocco, G. and Gaglio, G. (2013). Sociologie de l'innovation. *La nouvelle revue du travail*, 3, 26 October.

Florida, R. (2002). *The Rise of the Creative Class: And How It's Transforming Work, Leisure, Community and Everyday Life.* Basic Books, New York.

Foray, D. (2009). Introduction. *L'économie de la connaissance.* La Découverte, Paris.

Foray, D. and Cowan, R. (1998). *Economie de la codification et de la diffusion de la connaissance.* La Découverte, Paris.

Foray, D. and Lundvall, B.A. (1996). *Employment and Growth in the Knowledge-Based Economy.* OECD Paris.

Foray, D. and Lundvall, B.A. (1997). Une introduction à l'économie fondée sur la connaissance. *Économie de la connaissance et organisations*, 16–38.

Forest, J. and Hamdouch, A. (2009). Les clusters à l'ère de la mondialisation : fondements et perspectives de recherche. *Revue d'économie industrielle*, 128, 9–20, 15 December.

Forsé, M. and Degenne, A. (1997). *Les réseaux sociaux : une analyse structurale en sociologie*. Armand Colin, Paris.

Foucault, M. (2006). *La volonté de savoir : droit de mort et pouvoir sur la vie*. Folio, Paris.

Fouilleux, E. and Loconto, A.M. (2017). Dans les coulisses des labels : régulation tripartite et marchés imbriqués. De l'européanisation à la globalisation de l'agriculture biologique. *Revue française de sociologie*, 58(3), 501–531.

Fournet-Guérin, C. (2008). David Harvey. Géographie du capitalisme [Online]. Available at: https://laviedesidees.fr/Geographie-du-capitalisme.html.

Foyer, J. (2006). Développement des biotechnologies et biotechnologies pour le développement au Mexique : l'impossible appropriation. *Revue Tiers-Monde*, 188, 721–737.

Frances, J. (2012). Portrait du doctorant en entrepreneur. Les Doctoriales : un outil "au service de la professionnalisation des doctorants" ? *Mouvements*, 71(3), 54–65.

Gadille, M., Tremblay, D.G., Vion, A. (2013). La méta-organisation territorialisée, moteur d'apprentissages collectifs. *Revue Interventions économiques*, 48, 1 November.

Gaglio, G. (2011). *Sociologie de l'innovation*. Presses universitaires de France, Paris.

Galison, P. (1997). *Image and Logic: A Material Culture of Microphysics*. The University of Chicago Press, Chicago.

Garcia, S. (2009). Réformes de Bologne et économicisation de l'enseignement supérieur. *Revue du MAUSS*, 33, 154–172, 24 June.

Garnier, J.P. (1988a). Les technopoles : des métropoles de déséquilibre ? *Quaderni*, 6(1), 91–98.

Garnier, J.P. (1988b). Montpellier entre tekné et polis. In *Montpellier Europole*, Brunet, R. (ed.). GIP Reclus, Montpellier.

Garofoli, G. (1992). Les systèmes de petites entreprises : un cas paradigmatique de développement endogène. In *Les régions qui gagnent*, Lipietz, A. and Benko, G. (eds). Presses universitaires de France, Paris.

Gastaldi, L. and Lanciano-Morandat, C. (2017). L'enseignement supérieur et la recherche. Une pression temporelle accrue. *La nouvelle revue du travail*, 11, 31 October.

Gaudillière, J.P. (2002). *Inventer la biomédecine : la France, l'Amérique et la production des savoirs du vivant*. La Découverte, Paris.

Gaudillière, J.P. and Joly, P.B. (2006). Appropriation et régulation des innovations biotechnologiques : pour une comparaison transatlantique. *Sociologie du Travail*, 48, 330–349, 1 January.

Gaudillière, J.P. and Rheinberger, H.J. (2004). *From Molecular Genetics to Genomics: The Mapping Cultures of Twentieth-Century Genetics*. Routledge, London.

de Gaulejac, V. (2005). *La société malade de la gestion : idéologie gestionnaire, pouvoir managérial et harcèlement social*. Le Seuil, Paris.

de Gaulejac, V. and Hanique, F. (2015). *Le capitalisme paradoxant. Un système qui rend fou*. Le Seuil, Paris.

Gervais, P. (2015). Marchandisation ou managérialisation du savoir ? Réflexions sur le cas états-unien. In *La dérégulation universitaire. La construction étatisée des marchés universitaires dans le monde*, Christophe, C. and Soulié, C. (eds). Syllepses, Paris.

Gibbons, M., Limoges, C., Nowotny, H., Schwartzman, S., Scott, P., Trow, M. (1994). *The New Production of Knowledge: The Dynamics of Science and Research in Contemporary Societies*, 1st edition. Sage Publications, London.

Gilly, J.-P. and Grossetti, M. (1993). Organisation, individus et territoires : le cas des systèmes locaux d'innovation. *Revue d'économie régionale et urbaine*, 3, 449–468.

Gilly, J.-P. and Torre, A. (eds) (2000). *Dynamiques de proximité*. L'Harmattan, Paris.

Gingras, Y. and Roy, L. (2012). *Les universités nouvelles. Enjeux et perspectives*. Presses de l'Université du Québec, Montreal.

Giraudeau, M. (2007). Le travail entrepreneurial, ou l'entrepreneur schumpetérien performé. *Sociologie du Travail*, 49, 330–350, 1st January.

Godin, B. (2006). The linear model of innovation. The historical construction of an analytical framework. *Science, Tehnology and Human Values*, 31(6), 639–667.

Godin, B. and Gingras, Y. (2000). Impact of collaborative research on academic science. *Science and Public Policy*, 27(1), 65–73, 1 February.

Gollain, F. (2010). L'apport d'André Gorz au débat sur le capitalisme cognitif. *Revue du MAUSS*, 1(35), 541–558.

Gomez-Breysse, M. (2016). L'entrepreneur "lifestyle". *Revue de l'entrepreneuriat*, 15(3), 231–256, 20 December.

Gordon, I. and McCann, P. (2000). Industrial clusters: Complexes, agglomeration and/or social networks? *Urban Studies*, 37(3), 513–532, 1 March.

Gorz, A. (2003). *L'immatériel. Connaissance, valeur et capital*. Galilée, Paris.

Goussard, L. and Tiffon, G. (2013). Travailler en projets dans la R&D. Contraintes temporelles et transformations du travail de recherche. *Temporalités. Revue de sciences sociales et humaines*, 18, 19 December.

Granger, C. (2015). *La destruction de l'université française*. La Fabrique Éditions, Paris.

Granovetter, M.S. (1985). Economic action and social structure: The problem of embeddedness. *American Journal of Sociology*, 91(3), 481–510.

Granovetter, M.S. (2000). *Le marché autrement : recueil d'essais*. Desclée de Brouwer, Paris.

Grasland, L. (1988). Technopole et développement régional. In *Montpellier Europole*, Brunet, R. (ed.). GIP Reclus, Montpellier.

Grossetti, M. (1990). Enseignement supérieur et technopoles. Le cas de l'informatique à Toulouse. *Revue française de sociologie*, 31(3), 463–482.

Grossetti, M. (1995). *Science, industrie et territoire*. Presses universitaires du Mirail, Toulouse.

Grossetti, M. (2001). Genèse de deux systèmes urbains d'innovation en France : Grenoble et Toulouse. *Réalités industrielles. Annales des mines*, 68–72.

Grossetti, M. (2004). Concentration d'entreprises et innovation : esquisse d'une typologie des systèmes productifs locaux. Abstract. *Géographie, économie, société*, 6(2), 163–177.

Grossetti, M. (2006a). Les limites de la symétrie. À propos de l'ouvrage de Bruno Latour. Changer de société. Refaire de la sociologie. *Sociologies*, 1, 1–8.

Grossetti, M. (2006b). Réseaux sociaux et ressources de médiation dans l'activité économique. *Sciences de la société*, 73, 83–103.

Grossetti, M. (2009). Qu'est-ce qu'une relation sociale ? Un ensemble de médiations dyadiques. *Revista Hispana para el analis de redes sociales*, 6(2), 44–62.

Grossetti, M. and Bès, M.-P. (2001). Encastrements et découplages dans les relations science – industrie. *Revue française de sociologie*, 42(2), 327–355.

Grossetti, M. and Bès, M.-P. (2003). La dynamique des cercles et des réseaux. Encastrements et découplages. *Revue d'économie industrielle*, 103, 43–58.

Guillaume, R. (2008). Des systèmes productifs locaux aux pôles de compétitivité : approches conceptuelles et figures territoriales du développement. *Géographie, économie, société*, 10(3), 295–309, 22 September.

Guthleben, D. and Le Faou, O. (2011). *Une course pour la vie : l'AFM et la recherche biologique et médicale*. Armand Colin, Paris.

Haber, S. (2012). From Neoliberalism to Neocapitalism. Some reflections drawn from Foucault. *Actuel Marx*, 51, 59–72, 30 May.

Habermas, J. (1990). *La technique et la science comme "idéologie"*. Gallimard, Paris.

Habhab-Rave, S. (2010). Le rôle des communautés de pratique dans le processus de gestion des connaissances dans les entreprises innovantes : une étude de cas par comparaison intersites. *La Revue des sciences de gestion*, 241, 43–54, 1 November.

Häfner, S. (2010). Les iGEM 2009. La biologie synthétique. *Médecine/Sciences*, 26(11), 986–992, 1 November.

Hagstrom, W.O. (1965). *The Scientific Community*. Basic Books, New York.

Haila, A. (2008). The University of Heslinki as a developer. In *Global Universities and Urban Development. Case Studies and Analysis*, Wiewel, W., Perry, D.C. (eds). Routledge, Cambridge.

Hamdouch, A. (2007). Innovation clusters and networks: A critical review of the recent literature. *19th EAEPE Conference*, University of Porto.

Hamdouch, A. and Depret, M.H. (2009). Les clusters et les réseaux comme fondements de la dynamique d'innovation dans l'industrie biopharmaceutique. Working papers 8, Réseau de recherche sur l'innovation.

Harvey, D. (1989). From managerialism to entrepreneurialism: The transformation in urban governance in late capitalism. *Geografiska Annaler. Series B, Human Geography*, 71(1), 3–17.

Harvey, D. (2008). *Géographie de la domination*. Les Prairies ordinaires, Paris.

Hassenteufel, P. (2010). Les processus de mise sur agenda : sélection et construction des problèmes publics. *Informations Sociales*, 157, 50–58, 1 February.

Hatchuel, A. (1996). Comment penser l'action collective ? Théorie des mythes rationnels. *Annales littéraires de Besançon*, 7–43.

Hatchuel, A., Le Masson, P., Weil, B. (2002). De la gestion des connaissances aux organisations orientées conception. *Revue internationale des sciences sociales*, 171, 29–42.

Hatchuel, A., Le Masson P., Nakhla, M. (2004). *Plateformes techniques et politiques scientifiques : vers de nouvelles logiques de constitution et de gestion des dispositifs techniques partagés dans un contexte de recherche*. INRA Editions, Versailles.

Heil, C. (2010). De la science à la technoscience, du chercheur au technologue : les cas de la génopole d'Évry, du biopôle de Laval (Québec) et de l'Université de San Diego (Californie). Sociology Thesis, Université d'Évry Val d'Essonne, Évry.

Heil, C. and Lacroix, G. (2003). Une technopole atypique : le Genopole d'Évry. *Regards sociologiques sur les biotechnologies*, 79–108.

Helgason, S. (1997). International benchmarking experiences from OECD countries. Report, OECD, Copenhagen.

Hellec, F. (2014). Le rapport au terrain dans une thèse CIFRE. Du désenchantement à la distanciation. *Sociologies pratiques*, 28, 101–109.

Hirst, P. and Zeitlin, J. (1992). Flexible specialization versus post-Fordism: Theory, evidence and policy implications. *Pathways to Industrialization and Regional Development*, Storper, M. and Scott, A.J. (eds). Routledge, London/New York.

Houel, M. and Daunis, M. (2009). Les pôles de compétitivité : bilan et perspectives d'une politique industrielle et d'aménagement du territoire. Report, Sénat, Paris.

Hubert, M. and Louvel, S. (2012). Le financement sur projet : quelles conséquences sur le travail des chercheurs ? *Mouvements*, 71, 13–24, 22 September.

Hurel, F. (1999). *Start-up en France, des mythes aux réalités*. Étude ACPE, Paris.

Hussler, C. and Ronde, P. (2005). Innovation in regions: What does really matter? *Research Policy*, 24, 1150–1172.

d'Iribarne, P. (1993). *La logique de l'honneur. Gestion des entreprises et traditions nationales*. Le Seuil, Paris.

d'Iribarne, P. (1998). Cultures nationales et conceptions du travail. In *Traité de sociologie du travail*, 2nd edition, De Coster, M. and Pichault, F. (eds). De Boeck Supérieur, Paris.

Irvine, J. and Martin, B.R. (1984). *Foresight in Science: Picking the Winners*. Pinter Pub, London.

Jacq, A. (2011). L'OIN du plateau de Saclay. Science contre démocratie ou science sans démocratie ? [Online]. Available at: http://www.espaces-marx.net/IMG/pdf/OIN_Plateau_Saclay__2_.pdf.

Jasanoff, S. and Sang-Hyun, K. (2009). Containing the atom: Sociotechnical imaginaries and nuclear power in the United States and South Korea. *Minerva*, 47(2), 119–146.

Joliot, P. (2013). Recherche fondamentale et recherche appliquée. In *La mondialisation de la recherche : compétition, coopérations, restructurations*, Fussman, G. (ed.). Collège de France, Paris.

Joly, P.B. (1997). Chercheurs et laboratoires dans la nouvelle économie de la science. *Revue d'économie industrielle*, 79(1), 77–94.

Joly, P.B. (2015). Le régime des promesses technoscientifiques. In *Sciences et technologies émergentes : pourquoi tant de promesses ?* Audetat, M. (ed.). Hermann, Paris.

Karpik, L. (1972). Le capitalisme technologique. *Sociologie du travail*, 1, 2–34, January–March.

Kaufmann, A. (2004). Mapping the human genome at Genethon laboratory. French Muscular Distrophy Association and the politics of the gene. In *From Molecular Genetics to Genomic: The Mapping Cultures of Twentieth-Century Genetics*, Gaudillière, J.P. and Rheinberger, H.J. (eds). Routledge, London.

Keck, F. (2003). Des biotechnologies au biopouvoir, de la bioéthique aux biopolitiques. *Multitudes*, 12(2), 179–187.

Kingdon, J.W. (1984). *Agendas, Alternatives and Public Policies*. TBS The Book Service Ltd, Boston, MA.

Kleinman, D.L. and Vallas, S.P (2001). Sciences, capitalism, and the rise of the "Knowledge Worker": The changing structure of knowledge production in the United States. *Theory and Society*, 30(4), 451–492.

Kourtessi-Philippakis, G. (2011). La notion de territoire : définitions et approches. In *Archéologie du territoire, de l'Egée au Sahara*, Kourtessi-Philippakis, G. and Treuil, R. (eds). Publications de la Sorbonne, Paris.

Krugman, P.R. (1991). *Geography and Trade*. Leuven University Press/The MIT Press, Leuven/Cambridge.

Kuhn, T. (1990). *La tension essentielle. Tradition et changement dans les sciences*. Gallimard, Paris.

Lafontaine, C. (2014). *Le corps-marché. La marchandisation de la vie humaine à l'ère de la bioéconomie*. Le Seuil, Paris.

Lakomski-Laguerre, O. (2006). Introduction à Schumpeter. *L'Économie politique*, 29(1), 82–98.

Lallement, M. (2015). *L'Âge du faire. Hacking, travail, anarchie*. Le Seuil, Paris.

Lamy, E. (2017). La pensée gestionnaire de l'innovation et son enseignement. In *L'innovation en eaux troubles. Sciences, techniques, idéologies*, Sainsaulieu, I. and Saint-Martin, A. (eds). Éditions du Croquant, Vulaines-sur-Seine.

Lamy, J. and Le Roux, R. (2017). Situer l'innovation ? Le concept de cluster, outil performatif pour transformer la recherche publique. In *L'innovation en eaux troubles. Sciences, techniques, idéologies*, Sainsaulieu, I. and Saint-Martin, A. (eds). Éditions du Croquant, Vulaines-sur-Seine.

Lamy, J. and Saint-Martin, A. (2017). La sociologie de l'innovation scientifique et technique : un panorama historique et quelques hypothèses critiques. In *L'innovation en eaux troubles. Sciences, techniques, idéologies*, Sainsaulieu, I. and Saint-Martin, A. (eds). Éditions du Croquant, Vulaines-sur-Seine.

Lamy, E. and Shinn, T. (2006). L'autonomie scientifique face à la mercantilisation. *Actes de la recherche en sciences sociales*, 164(4), 23–50.

Lanciano-Morandat, C., Jolivet, E., Gurnay, T., Nohara, H., Pardo, D. (2009). Le capital social des entrepreneurs comme indice de l'émergence de cluster ? *Revue d'économie industrielle*, 128, 177–202, 23 December.

Laperche, B. and Uzunidis, D. (2010). La territorialisation de la recherche et de l'innovation. Orientation, programmation et coordination. *Cahiers du Lab. RII*, 110, 1–22.

Larédo, P. and Mustar, P. (2004). La recherche publique en France : évolutions et enjeux. *Le Banquet*, 19, 95–113.

Largier, A., Lartigue, S., Soulard, O., Tarquis, C. (2008). Clusters mondiaux : regards croisés sur la théorie et la réalité des clusters. Identification et cartographie des principaux clusters internationaux. Report, Institut d'aménagement et d'urbanisme de la Région Île-de-France, Paris.

Lascoumes, P. and Le Galès, P. (2005). *Gouverner par les instruments*. Les Presses de Sciences Po, Paris.

Latour, B. (1989). *La science en action : introduction à la sociologie des sciences*. La Découverte, Paris.

Latour, B. and Woolgar, S. (2006). *La vie de laboratoire*. La Découverte, Paris.

Laurent, C. and Du Tertre, C. (2008). *Secteurs et territoires dans les régulations émergentes*. L'Harmattan, Paris.

Lazega, E. (2014). *Réseaux sociaux et structures relationnelles*, 3rd edition. Presses universitaires de France, Paris.

Le Rudulier, K. (2016). Premiers retours sur le statut d'étudiant-entrepreneur. *Informations sociales*, 195, 135–143.

Leducq, D. and Lusso, B. (2011). Le cluster innovant : conceptualisation et application territoriale. *Cybergeo: European Journal of Geography*, 7 March.

Levine, M. (2004). La "classe créative" et la prospérité urbaine : mythes et réalités. *Villes, Régions, Monde*. INRS Urbanisation, Montreal.

Levy, R. (2005). Les doctorants CIFRE : médiateurs entre laboratoires de recherche universitaires et entreprises. *Revue d'économie industrielle*, 111(1), 79–96.

Levy, R. and Talbot, D. (2015). Control by proximity: Evidence from the "Aerospace Valley" competitiveness cluster. *Regional Studies*, 49(6), 955–972.

Leydesdorff, L. and Etzkowitz, H. (2000). Le "Mode 2" et la globalisation des systèmes d'innovation "nationaux". *Sociologie et sociétés*, 32, 135–156.

Liefooghe, C., Delmer, S., Prevot, M., Stachowski, K. (2006). De Villeneuve-d'Ascq Technopole à la métropole technopolitaine : espace, réseaux, gouvernance. *Norois*, 200, 51–76.

Linhart, D. (2010). *La modernisation des entreprises*, 3rd edition. La Découverte, Paris.

Linhart, D. (2015). *La comédie humaine du travail : de la déshumanisation taylorienne à la sur-humanisation managériale*. Erès, Toulouse.

Lipinski, M. (2013). *Sciences, un enjeu citoyen. Une politique écologiste de la recherche et de l'innovation : une politique écologiste de la recherche et de l'innovation*. Les Petits Matins, Paris.

Loeve, S. (2017). La loi de Moore : enquête critique sur l'économie d'une promesse. In *L'innovation en eaux troubles. Sciences, techniques, idéologies*, Sainsaulieu, I. and Saint-Martin, A. (eds). Éditions du Croquant, Vulaines-sur-Seine.

Lordon, F. (2000). La force des idées simples. Misère épistémique des comportements économiques. *Politix*, 52, 183–209.

Louvel, S. (2005). La construction locale des laboratoires. Approche ethnographique de dynamiques d'évolution de laboratoires académiques en France. Sociology PhD Thesis, Université Pierre Mendès-France, Grenoble.

Louvel, S. (2006). Les doctorants en sciences expérimentales : futurs collègues ou jeunes collègues ? *Formation emploi. Revue française de sciences sociales*, 96, 53–66, 1 October.

Louvel, S. and Hubert, M. (2016). L'usage des exemples étrangers dans les politiques de financement de la recherche. Les nanosciences et nanotechnologies en France. *Revue française de sociologie*, 57(3), 473–501.

Lundvall, B.A. and Archibuji, D. (2001). *The Globalizing Learning Economy*. Oxford University Press, Oxford.

Lundvall, B.A. and Borras, S. (1997). The globalising learning economy: Implications for innovation policy. Report, European Commission, Brussels.

Lusso, B. (2013). Pôles transmédia, actions d'intermédiation et construction de liens au sein de l'industrie du multimédia des aires métropolitaines de Lille, Lyon et Marseille. *Revue Interventions économiques*, 48.

Machlup, F. (1962). *The Production and Distribution of Knowledge in the United States*. Princeton University Press, Princeton, NJ.

Maestrutti, M. and Kröger, F. (2018). *Les imaginaires et les techniques*, 1st edition. Presses des Mines, Paris.

Maillat, D., Crevoisier, O., Vasserot, J.Y. (1992). Innovation et district industriel : l'Arc jurassien suisse. In *Entreprises innovatrices et développement territorial*, Maillat, D. and Perrin, J.C. (eds). GREMI, EDES, Neuchâtel.

Malissard, P., Gingras, Y., Gemme, B. (2003). La commercialisation de la recherche. *Actes de la recherche en sciences sociales*, 148(1), 57–67.

Mangematin, V. (2003). Les docteurs, producteurs et diffuseurs de connaissances. In *Encyclopédie de l'innovation*, Mustar, P. and Penan, H. (eds). Economica, Paris.

Mangematin, V. and Peerbaye, A. (2004). Les grands équipements en sciences de la vie : quelle politique publique ? *Revue française d'administration publique*, 112(4), 705–718.

Marshall, A. (1990). Organisation industrielle : la concentration d'industries spécialisées dans certaines localités. *Revue française d'économie*, 5(3), 155–170.

Marshall, A. (2010). *Principles of Economics*. Nabu Press.

Martin, R. and Sunley, P. (2003). Deconstructing clusters: Chaotic concept or policy panacea? *Journal of Economic Geography*, 3(1), 5–35, 1 January.

Marty, O., Moreau, R., Weinberger, D. (2004). *Start-up ? Du mythe médiatique aux réalités sociologiques*. L'Harmattan, Paris.

Marx, K. (1867). Chapitre V : Procès de travail et procès de valorisation. *Le Capital. Livre I*. Presses universitaires de France, Paris.

Mazzucato, M. (2015). *The Entrepreneurial State: Debunking Public vs. Private Sector Myths*, 2nd edition. Public Affairs, New York.

Menger, P.-M. (2003). *Portrait de l'artiste en travailleur : métamorphoses du capitalisme*. Le Seuil, Paris.

Menger, P.-M. (2009). *Le travail créateur. S'accomplir dans l'incertain*. Le Seuil, Paris.

Menger, P.-M. (2015). La production du savoir. Carrières et compétitions dans l'enseignement et la recherche. Cours Amphithéâtre Marguerite de Navarre. Marcelin Berthelot, Collège de France, Paris.

Merton, R.K. (1996). *On Social Structure and Science*, 1st edition. University of Chicago Press, Chicago, IL.

MESR (2013). L'état de l'emploi scientifique en France. Report, Ministère de l'Enseignement supérieur, de la Recherche et de l'Innovation, Paris.

Metcalfe, J.S. (1995). The economic foundations of technology policies: Equilibrium and evolutionary perspectives. In *Handbook of the Economics of Innovation and Technological Change*, Stoneman, P. (ed.). Blackwell, Oxford.

Métivier, F., Lemaire, P., Riot, E. (2015). CIR et R&D : efficacité du dispositif depuis la réforme de 2008. Report, Association de chercheurs Sciences en marche, Paris, 6 April.

Meulders, D., O'Dorchai, S., Simeu, N. (2012). Les inégalités entre femmes et hommes dans les universités francophones de Belgique. GENIUF research report, Université libre de Bruxelles, Brussels.

Milgram, S. (1967). The small world problem. *Psychology Today*, 2, 60–67.

de Montlibert, C. (2012). Les apories des discours sur la mondialisation : du mythe aux réalités. *Regards sociologiques*, 43–44, 13–25.

Morel-Journel, C. and Sala-Pala, V. (2011). Le peuplement, catégorie montante des politiques urbaines néolibérales ? Le cas de Saint-Étienne. *Métropoles*, 10.

Morin, E. (1958). *Le cinéma ou l'homme imaginaire*. Éditions de Minuit, Paris.

Moulier, B.Y. (2007). *Le capitalisme cognitif : la nouvelle grande transformation*. Éditions Amsterdam, Paris.

Moulier, B.Y. (2010). *L'abeille et l'économiste*. Carnets Nord, Paris.

Musselin, C. (2008). Vers un marché international de l'enseignement supérieur ? *Critique internationale*, 39, 13–24, 26 June.

Mustar, P. and Laredo, P. (2002). Innovation and research policy in France (1980–2000) or the disappearance of the Colbertist state. *Research Policy*, 31(1), 55–72.

Nakara, W.A. and Fayolle, A. (2012). Les “bad” pratiques d’accompagnement à la création d’entreprise. *Revue française de gestion*, 228–229, 231–251.

Nalebuff, B. and Brandenburger, A. (1996). *La co-opétition, une révolution dans la manière de jouer concurrence et coopération*. Village Mondial, Paris.

Novarina, G. (2012). L’ancrage territorial de l’économie : du district industriel à la métropole. *L’Économie politique*, 53, 16–25, 28 February.

Nowotny, H., Scott, P., Gibbons, M. (2003). *Repenser la science. Savoir et société à l’ère de l’incertitude*. Belin, Paris.

OECD (1996). The knowledge-based economy. Report, Organisation de coopération et de développement économique, Paris.

OECD (2006). Science, technologie et industrie. Perspectives de l’OCDE 2006 : principales conclusions. Report, Organisation de coopération et de développement économique, Paris.

Olivier de Sardan, J.P. (1995). La politique du terrain. *Enquête. Archives de la revue Enquête*, 1, 71–109, 1 October.

Owen-Smith, J. and Powell, W.W. (2003). The expanding role of university patenting in the life sciences: Assessing the importance of experience and connectivity. *Research Policy*, 32(9), 1695–1711.

Owen-Smith, J. and Powell, W.W. (2004). Knowledge networks as channels and conduits: The effects of spillovers in the Boston biotechnology community. *Organization Science*, 15(1), 5–21, 1 February.

Paillard, S. (2001). Économie de la connaissance. Document de synthèse des travaux du séminaire d’experts. Report, Commissariat général du plan, Paris.

Panese, F. (2017). Cerveau et imaginaire sociotechnique : genèse du Human Brain Project entre science et politique. In *L’innovation en eaux troubles. Sciences, techniques, idéologies*, Sainsaulieu, I., Saint-Martin, A. (eds). Éditions du Croquant, Vulaines-sur-Seine.

Pecqueur, B. (1995). *Les relations entre les entreprises industrielles du sillon alpin*. IRERD, Grenoble.

Pecqueur, B. and Zimmermann, J.B. (eds) (2004). *Économie de Proximités*. Hermès, Lavoisier, Paris.

Peerbaye, A. (2004). La Construction de l’espace génomique en France : la place des dispositifs instrumentaux. Sociology PhD Thesis, École normale supérieure, Paris.

Perkmann, M. and Walsh, K. (2007). University – Industry relationships and open innovation: Towards a research agenda. *International Journal of Management Reviews*, 9(4), 259–280, 1 December.

Perrin-Joly, C. (2010). De la recherche salariée en France : lien de subordination et liberté de la recherche. *Sociologies* [Online]. Available at: http://journals.openedition.org/sociologies/3380.

Pestre, D. (1994). Le renouveau de la recherche à l'École polytechnique et le laboratoire de Louis Leprince-Ringuet. In *La formation polytechnicienne 1794–1994*, Belhoste, B., Dahan Dalmedico, A., Picon, A. (eds). Dunod, Malakoff.

Petit, V. and Mari, I. (2009). *La légitimité des équipes dirigeantes : une dimension négligée de la gouvernance d'entreprise*. Cahier de recherche de l'Edhec.

Pfister, M. and Nemeti, F. (1995). District de technologique et innovation : le cas de l'Arc jurassien suisse. *Géocarrefour*, 70(1), 27–34.

Pichard-Stamford, J.P. (2000). Légitimité et enracinement du dirigeant par le réseau des administrateurs. *Finance Contrôle Stratégie*, 3(4), 143–178.

Piore, M. and Sabel, C. (1986). *The Second Industrial Divide: Possibilities for Prosperity*, Reissue. Basic Books, New York.

Piotet, F. and Sainsaulieu, R. (1994). *Méthodes pour une sociologie de l'entreprise.* Presses de la fondation nationale des sciences politiques, Paris.

Piron, S. (2008). Lisons Peter Lawrence, ou les implications morales de l'évaluation bibliométrique. *Evaluation de la recherche en SHS* [Online]. Available at: http://evaluation.hypotheses.org/229.

Pollock, N. and Williams, R. (2010). The business of expectations: How promissory organizations shape technology and innovation. *Social Studies of Science*, 40(4), 525–548.

Popovic, P. (2008). *Imaginaire social et folie littéraire. Le second Empire de Paulin Gagne*. Les Presses de l'Université de Montréal, Montreal.

Porter, M. (1990). *Competitive Advantage of Nations*. Free Press, New York.

Porter, M. (1993). *L'avantage concurrentiel des nations*. Éditions du Renouveau pédagogique, Montreal.

Porter, M. (1998a). Clusters and competition: New agendas for companies, governments, and institutions. Harvard Business School Working Paper, Cambridge, MA, 98/080.

Porter, M. (1998b). Clusters and the new economics of competition. *Harvard Business Review*, 76(6), 77–90.

Porter, M. (1998c). *On Competition*. Harvard Business School Press, Boston, MA.

Porter, M. (2000). Locations, clusters and company strategy. In *The Oxford Handbook of Economic Geography*, Clark, G., Feldman, M., Gertler, M. (eds). Oxford University Press, Oxford.

Portes, A. (1995). *The Economic Sociology of Immigration: Essays on Networks, Ethnicity and Entreoreneurship*. Russel Sage Foundation, New York.

Prades, J. (1992). *La technoscience. Les fractures du discours*. L'Harmattan, Paris.

Quenson, E. (2012). La formation en entreprise : évolution des problématiques de recherche et des connaissances. *Savoirs*, 28, 11–63, 1 October.

Quet, M. (2012). La critique des technologies émergentes face à la communication promettante. Contestations autour des nanotechnologies. *Réseaux*, 173/174, 271–302, 25 July.

Rabinow, P. (2000). *Le déchiffrage du génome : l'aventure française*. Odile Jacob, Paris.

Rallet, A. and Torre, A. (eds) (1995). *Économie industrielle et économie spatiale*. Economica, Paris.

Rameaux, C. and Ribac, F. (2003). À plusieurs voix sur Portrait de l'artiste en travailleur. *Mouvements*, 29(4), 146–154.

Rasse, P. and Arasziewiez, J. (2007). *Les technopoles entre utopie et non-lieu de la mondialisation : le cas de Sophia Antipolis*. Presses universitaires de Bordeaux, Bordeaux.

Rave-Habhab, S. (2012). Place de la culture d'entreprise dans un processus d'innovation chez les PME de haute technologie : une étude de cas par comparaison inter sites. *Management et Avenir*, 50, 78–104, 10 April.

Ravelli, Q. and Rosental, C. (2007). Les capitalistes de la science. Enquête sur les démonstrateurs de la Silicon Valley et de la NASA. *Lectures*, 12 November.

Raveyre, M.F. and Saglio, J. (1984). Les systèmes industriels localisés : éléments pour une analyse sociologique des ensembles de PME industriels. *Sociologie du travail*, 26(2), 157–176.

Ray, P.H. and Anderson, S.R. (2001). *L'émergence des créatifs culturels*. Yves Michel, Barret-le-Bas.

Reich, R. (1997). *L'économie mondialisée*. Dunod, Paris.

Renaud, C. (2015). Étude des rapports locaux entre science, industrie et pouvoirs publics : les conditions socioculturelles, économiques et techniques de l'engagement partenarial au sein d'un pôle de compétitivité. Sociology PhD Thesis, Bordeaux.

Rifkin, J. (1997). *La fin du travail*. La Découverte, Paris.

Rifkin, J. (2012). *La troisième révolution industrielle*. Les Liens qui Libèrent, Paris.

Rip, A. (2002). Regional innovation system and the advent of strategic science. *Technology Transfer*, 27, 123–131.

Rip, A. (2015). L'avenir du régime des promesses. In *Sciences et technologies émergentes : pourquoi tant de promesses ?* Audetat, M. (ed.). Hermann, Paris.

Rip, A. and Van Lente, H. (1998). The rise of membrane technology: From rhetoric to social reality. *Social Studies of Science*, 28(2), 221–254.

Robelet, M. (2016). La guerre des mondes associatifs aura-t-elle lieu ? In *Faire la concurrence. Retour sur un phénomène social et économique*, Castel, P., Hénaut, L., Marchal, E. (eds). Presses des Mines, Paris.

Rogers, E.M. (2003). *Diffusion of Innovations*, 5th edition. Free Press, New York.

Rosa, H. (2010). *Accélération*. La Découverte, Paris.

Rosenberg, N. (1974). Karl Marx on the economic role of science. *Journal of Political Economy*, 82(4), 713–728.

Rosenberg, N. (1983). *Inside the Black Box: Technology and Economics*. Cambridge University Press, Cambridge.

Rosental, C. (2007). *Les capitalistes de la science : enquête sur les démonstrateurs de la Silicon Valley et de la NASA*. CNRS, Paris.

Rosental, C. (2017). Les conditions sociales des échanges dans la Silicon Valley. *Zilsel*, 1, 55–81, 15 February.

Sahlin, K. and Wedlin, L. (2008). Circulating ideas: Imitation, translation and editing. In *The SAGE Handbook of Organizational Institutionalism*, Greenwood, R., Oliver, C., Sahlin, K., Suddaby, R. (eds). SAGE Publications, London.

Sainsaulieu, R. (1988). *L'identité au travail. Les effets culturels de l'organisation*. Presses de Sciences Po, Paris.

Sainsaulieu, I. and Saint-Martin, A. (2017). *L'innovation en eau trouble : sciences, techniques et idéologies*. Éditions du Croquant, Vulaines sur Seine.

de Saint-Georges, P. (1993). Culture d'entreprise, communication interne et stratégies de changement. *Communication et organisation* [Online]. Available at: http://journals.openedition.org/communicationorganisation/1657.

Sassen, S. (2009). *La globalisation. Une sociologie*. Gallimard, Paris.

Savall, H. and Zardet, V. (2005). *Ingénierie stratégique du roseau*, 2nd edition. Economica, Paris.

Saxenian, A. (1996). *Regional Advantage: Culture and Competition in Silicon Valley and Route 128*, 5th edition. Harvard University Press, Cambridge, MA.

Saxenian, A. (2000). Silicon Valley : les secrets d'une réussite. *Sciences humaines. Hors série : Les nouveaux visages du capitalisme*, 29.

Schultheis, F., Roca i Escoda, M., Cousin, P.F. (2008). *Le Cauchemar de Humboldt. Les réformes de l'enseignement supérieur européen Franz.* Raisons d'agir, Paris.

Schumpeter, J. (1934). *The Theory of Economic Development: An Inquiry Into Profits, Capital, Credit, Interest, and the Business Cycle*. Transaction Publishers, New Brunswick.

Schumpeter, J. (1999). *Théorie de l'évolution économique. Recherches sur le profit, le crédit, l'intérêt*, Reissue. Dalloz-Sirey, Paris.

Schwartz, O. (1993). L'empirisme irréductible. *Le hobo, sociologie du sans-abri*. Armand Colin, Paris.

Schweisguth, E. (1981). Shinn Terry, l'École polytechnique. *Revue française de sociologie*, 22/4, 655–657.

Sedel, J. (2009). *Les médias et la banlieue*. Le Bord de l'eau, Paris.

Sheila, J. and Kim, S.-H. (2015). *Dreamscapes of Modernity. Sociotechnical Imaginaries and the Fabrication of Power*. University of Chicago Press, Chicago, IL.

Shinn, T. (2002). Nouvelle production du savoir et triple hélice [Tendances du prêt-à-penser les sciences]. *Actes de la recherche en sciences sociales*, 141(1), 21–30.

Shinn, T. (2007). *Research-Technology and Cultural Change: Instrumentation, Genericity, Transversality*. The Bardwell Press, Oxford.

Simoulin, V. (2012). *Sociologie d'un grand équipement scientifique. Le premier synchrotron de troisième génération*. ENS Éditions, Lyon.

Sintez, C. (2012). De la subordination à la domination du salarié : la "preuve sociologique" par le temps de travail. *Siècles*, 35–36 [Online]. Available at: http://journals.openedition.org/siecles/1431 [Accessed 2 March 2021].

Slaughter, S. (1997). *Academic Capitalism: Politics, Policies, and the Entrepreneurial University*. Johns Hopkins University Press, Baltimore, MD.

Slaughter, S. and Rhoades, G. (2010). *Academic Capitalism and the New Economy: Markets, State, and Higher Education*. Johns Hopkins University Press, Baltimore, MD.

Star, L. and Griesemer, J. (1989). Institutional ecology, "Translations", and Boundary objects: Amateurs and professionals on Berkeley's museum of vertrebate zoologie. *Social Studies of Science*, 19(3), 387–420.

Stehr, N. (2000). Le savoir en tant que pouvoir d'action. *Sociologie et sociétés*, 32(1), 157–170.

Stephan, P. (2015). *How Economics Shapes Science*, Reissue. Harvard University Press, Cambridge, MA.

Stevanovic, B. (2012). Orientations scientifiques des filles en France : un bilan contrasté. *Questions vives*, 6(16).

Storper, M. and Scott, A. (1990). Work organisation and local labour markets in an era of flexible production. *International Labour Review*, 129, 573–591.

Suire, R. and Vicente, J. (2015). Récents enseignements de la théorie des réseaux en faveur de la politique et du management des clusters. *Revue d'économie industrielle*, 152, 91–119.

Sunder-Rajan, K. (2006). *Biocapital: The Constitution of Postgenomic Life*. Duke University Press, Durham/London.

Talbot, D. and Kirat, T. (2005). Proximité et institutions : nouveaux éclairages. *Économie et institutions*, 6/7, 9–15, 1 December.

Tanguy, L. (2016). *Enseigner l'esprit d'entreprise à l'école : le tournant politique des années 1980–2000 en France*. La Dispute, Paris.

Tatsuno, S. (1987). *Les technopoles ou la révolution de l'intelligence*. Les Éditions d'Organisation, Paris.

Thébaud-Mony, A. (2014). *La science asservie*. La Découverte, Paris.

Tinel, B. (2006). Karl Marx : l'organisation et l'exploitation du travail. In *Encyclopédie des ressources humaines*, Allouche, J. (ed.), 2nd edition. Vuibert, Paris.

Torre, A. (2006). Clusters et systèmes locaux d'innovation. Retour critique sur les hypothèses naturalistes de transmission des connaissances à l'aide des catégories de l'économie de la proximité. *Région et Développement*, 24, 15–43.

Torre, A. and Filippi, M. (eds) (2005). *Proximités et changements socio-économiques dans les mondes ruraux*. Inra Editions, Paris.

Torre, A. and Zuindeau, B. (2012). Les apports de l'économie de la proximité aux approches environnementales : inventaire et perspectives. *Natures Sciences Sociétés*, 17(4), 349–360, 1 January.

Touraine, A. (1969). *La société post-industrielle*. Denoël, Paris.

Toutain, O. and Verzat, C. (2017). L'entrepreneuriat et la jeunesse, un sujet en quête de sens. *Entreprendre et Innover*, 33, 5–9, 27 November.

Trépanier, M. and Ippersiel, M.P. (2003). Hiérarchie de la crédibilité et autonomie de la recherche. Summary, Zusammenfassung, Resumen. *Actes de la recherche en sciences sociales*, 148, 74–82.

Uhalde, M. (2008). L'instrumentalisation de la sociologie en situation d'intervention : analyse critique d'une notion ordinaire. *Sociologies pratiques*, 16, 95–113.

Uzunidis, D. (2003). Les facteurs actuels qui font de la science une force productive au service du capital. Le quatrième moment de l'organisation de la production. *Innovations*, 17(1), 51–78.

Vallier, E. (2011). L'application du processus de Bologne dans les universités espagnoles : entre influence européenne, rôle de l'État et revendications étudiantes. Political Science Master's Thesis, Université Lumière Lyon II and Université Complutense de Madrid, Lyon.

Vallier, E. (2018). Voyage en cluster, qu'en est-il de sa promesse relationnelle ? *ARCS – Analyse de réseaux pour les sciences sociales*, March.

Van Lente, H. (1993). *Promising Technology. The Dynamics of Expectations in Technological Developments*. Twente University, Enschede, The Netherlands.

Vatin, F. (2015). *Le Travail et ses valeurs*. Albin Michel, Paris.

Veltz, P. (1996). *Mondialisation, villes et territoires. L'économie d'archipel*. Presses universitaires de France, Paris.

Vercellone, C. (2003). *Sommes-nous sortis du capitalisme industriel ?* La Dispute, Paris.

Vercellone, C. (2008). La thèse du capitalisme cognitif. Une mise en perspective historique et théorique. In *Les nouveaux horizons du capitalisme*, Colletis, G., Paulré, B. (eds). Economica, Paris.

Vergès, E. (2010). La loi sur l'innovation et la recherche, une révolution douce du droit de la recherche ? In *L'innovation et la recherche en France, analyse juridique*, Robin, A. (ed.). Larcier, Brussels.

Vergès, E. (2014). Normes de la recherche scientifique. La valorisation sous tension : combats politiques et conflits de paradigmes autour de l'usage des résultats de la recherche. *Cahiers Droit, Sciences et Technologies*, 4, 243–250, 1 December.

Verzat, C. (2015). “Esprit d’entreprendre, es-tu là ?” Mais de quoi parle-t-on ? *Entreprendre et Innover*, 27, 81–92.

Vinck, D. and Zarama, G. (2007). La fusion de laboratoires. *Revue d’anthropologie des connaissances*, 1(2), 267–296, 1 July.

Vivant, E. (2006). La classe créative existe-t-elle ? Discussion des thèses de Richard Florida. *Les Annales de la recherche urbaine*, 101, 155–161.

Vuarin, R. (1994). L’argent et l’entregent. *Cahiers des sciences humaines*, 30(1–2), 255–273.

Watier, P. (2009). De la société aux formes de socialisation. *Sociétés*, 101, 49–61, 5 February.

Weber, M. (2003). *Économie et société, tome I : Les catégories de la sociologie.* Pocket, Paris.

Wenger, E. (2005). *La théorie des communautés de pratique. Apprentissage, sens et identité.* Les Presses de l’Université de Laval, Quebec.

Wheeler, C.H. (2001). Search, sorting, and urban agglomeration. *Journal of Labor Economics*, 19(4), 879–899, 1 October.

White, H. (1993). *Identity and Control: A Structural Theory of Social Action.* Princeton University Press, Princeton, NJ.

Winkin, Y. (2008). L’esprit de Bologne : “Si les universités ne s’adaptent pas, on se passera d’elles”. In *Le cauchemar de Humboldt : les réformes de l’enseignement supérieur européen*, Schultheis, F., Roca i Escoda, M., Cousin, P.F. (eds). Raisons d’agir, Paris.

Yacoub, N. and Laperche, B. (2010). Stratégies des grandes firmes pharmaceutiques face aux médicaments génériques. *Innovations*, 32, 81–107, 18 June.

Younes, D. (2011). Créer la coopération ? Les dynamiques de partenariat sur le pôle de compétitivité du plateau de Saclay. Sociology PhD Thesis, Institut d’études politiques de Paris, Paris.

Zarka, Y.C. (2016). La marchandisation du corps humain : la réalité du corps fétichisé. *Cités*, 65, 33–40.

Zepf, M. (2011). *Enjeux de la planification territoriale en Europe.* Presses polytechniques et universitaires romandes, Lausanne.

Index

Other titles from

ISTE

in

Innovation, Entrepreneurship and Management

2021

ARCADE Jacques
Strategic Engineering (Innovation and Technology Set – Volume 11)

BÉRANGER Jérôme, RIZOULIÈRES Roland
The Digital Revolution in Health (Health and Innovation Set – Volume 2)

BOBILLIER CHAUMON Marc-Eric
Digital Transformations in the Challenge of Activity and Work: Understanding and Supporting Technological Changes (Technological Changes and Human Resources Set – Volume 3)

BUCLET Nicolas
Territorial Ecology and Socio-ecological Transition (Smart Innovation Set – Volume 34)

DIMOTIKALIS Yannis, KARAGRIGORIOU Alex, PARPOULA Christina, SKIADIS Christos H
Applied Modeling Techniques and Data Analysis 1: Computational Data Analysis Methods and Tools (Big Data, Artificial Intelligence and Data Analysis Set - Volume 7)
Applied Modeling Techniques and Data Analysis 2: Financial, Demographic, Stochastic and Statistical Models and Methods (Big Data, Artificial Intelligence and Data Analysis Set – Volume 8)

DISPAS Christophe, KAYANAKIS Georges, SERVEL Nicolas, STRIUKOVA Ludmila
Innovation and Financial Markets (Innovation between Risk and Reward Set – Volume 7)

GOGLIN Christian
Emotions and Values in Equity Crowdfunding Investment Choices 2: Modeling and Empirical Study

GRENIER Corinne, OIRY Ewan
Altering Frontiers: Organizational Innovations in Healthcare (Health and Innovation Set – Volume 1)

GUERRIER Claudine
Security and Its Challenges in the 21st Century (Innovation and Technology Set – Volume 12)

HELLER David
Performance of Valuation Methods in Financial Transactions (Modern Finance, Management Innovation and Economic Growth Set – Volume 4)

LEHMANN Paul-Jacques
Liberalism and Capitalism Today

UZUNIDIS Dimitri, KASMI Fedoua, ADATTO Laurent
Innovation Economics, Engineering and Management Handbook 1: Main Themes
Innovation Economics, Engineering and Management Handbook 2: Special Themes

2020

ACH Yves-Alain, RMADI-SAÏD Sandra
Financial Information and Brand Value: Reflections, Challenges and Limitations

ANDREOSSO-O'CALLAGHAN Bernadette, DZEVER Sam, JAUSSAUD Jacques, TAYLOR Robert
Sustainable Development and Energy Transition in Europe and Asia (Innovation and Technology Set – Volume 9)

BEN SLIMANE Sonia, M'HENNI Hatem
Entrepreneurship and Development: Realities and Future Prospects (Smart Innovation Set – Volume 30)

CHOUTEAU Marianne, FOREST Joëlle, NGUYEN Céline
Innovation for Society: The P.S.I. Approach (Smart Innovation Set – Volume 28)

CORON Clotilde
Quantifying Human Resources: Uses and Analysis (Technological Changes and Human Resources Set – Volume 2)

CORON Clotilde, GILBERT Patrick
Technological Change (Technological Changes and Human Resources Set – Volume 1)

CERDIN Jean-Luc, PERETTI Jean-Marie
The Success of Apprenticeships: Views of Stakeholders on Training and Learning (Human Resources Management Set – Volume 3)

DELCHET-COCHET Karen
Circular Economy: From Waste Reduction to Value Creation (Economic Growth Set – Volume 2)

DIDAY Edwin, GUAN Rong, SAPORTA Gilbert, WANG Huiwen
Advances in Data Science (Big Data, Artificial Intelligence and Data Analysis Set – Volume 4)

DOS SANTOS PAULINO Victor
Innovation Trends in the Space Industry
(Smart Innovation Set – Volume 25)

GASMI Nacer
Corporate Innovation Strategies: Corporate Social Responsibility and Shared Value Creation
(Smart Innovation Set – Volume 33)

GOGLIN Christian
Emotions and Values in Equity Crowdfunding Investment Choices 1: Transdisciplinary Theoretical Approach

GUILHON Bernard
Venture Capital and the Financing of Innovation
(Innovation Between Risk and Reward Set – Volume 6)

LATOUCHE Pascal
Open Innovation: Human Set-up
(Innovation and Technology Set – Volume 10)

LIMA Marcos
Entrepreneurship and Innovation Education: Frameworks and Tools
(Smart Innovation Set – Volume 32)

MACHADO Carolina, DAVIM J. Paulo
Sustainable Management for Managers and Engineers

MAKRIDES Andreas, KARAGRIGORIOU Alex, SKIADAS Christos H.
Data Analysis and Applications 3: Computational, Classification, Financial, Statistical and Stochastic Methods
(Big Data, Artificial Intelligence and Data Analysis Set – Volume 5)
Data Analysis and Applications 4: Financial Data Analysis and Methods
(Big Data, Artificial Intelligence and Data Analysis Set – Volume 6)

MASSOTTE Pierre, CORSI Patrick
Complex Decision-Making in Economy and Finance

MEUNIER François-Xavier
Dual Innovation Systems: Concepts, Tools and Methods
(Smart Innovation Set – Volume 31)

MICHAUD Thomas
Science Fiction and Innovation Design (Innovation in Engineering and Technology Set – Volume 6)

MONINO Jean-Louis
Data Control: Major Challenge for the Digital Society (Smart Innovation Set – Volume 29)

MORLAT Clément
Sustainable Productive System: Eco-development versus Sustainable Development (Smart Innovation Set – Volume 26)

SAULAIS Pierre, ERMINE Jean-Louis
Knowledge Management in Innovative Companies 2: Understanding and Deploying a KM Plan within a Learning Organization (Smart Innovation Set – Volume 27)

2019

AMENDOLA Mario, GAFFARD Jean-Luc
Disorder and Public Concern Around Globalization

BARBAROUX Pierre
Disruptive Technology and Defence Innovation Ecosystems (Innovation in Engineering and Technology Set – Volume 5)

DOU Henri, JUILLET Alain, CLERC Philippe
Strategic Intelligence for the Future 1: A New Strategic and Operational Approach
Strategic Intelligence for the Future 2: A New Information Function Approach

FRIKHA Azza
Measurement in Marketing: Operationalization of Latent Constructs

FRIMOUSSE Soufyane
Innovation and Agility in the Digital Age (Human Resources Management Set – Volume 2)

GAY Claudine, SZOSTAK Bérangère L.
Innovation and Creativity in SMEs: Challenges, Evolutions and Prospects (Smart Innovation Set – Volume 21)

GORIA Stéphane, HUMBERT Pierre, ROUSSEL Benoît
Information, Knowledge and Agile Creativity (Smart Innovation Set – Volume 22)

HELLER David
Investment Decision-making Using Optional Models (Economic Growth Set – Volume 2)

HELLER David, DE CHADIRAC Sylvain, HALAOUI Lana, JOUVET Camille
The Emergence of Start-ups (Economic Growth Set – Volume 1)

HÉRAUD Jean-Alain, KERR Fiona, BURGER-HELMCHEN Thierry
Creative Management of Complex Systems (Smart Innovation Set – Volume 19)

LATOUCHE Pascal
Open Innovation: Corporate Incubator (Innovation and Technology Set – Volume 7)

LEHMANN Paul-Jacques
The Future of the Euro Currency

LEIGNEL Jean-Louis, MÉNAGER Emmanuel, YABLONSKY Serge
Sustainable Enterprise Performance: A Comprehensive Evaluation Method

LIÈVRE Pascal, AUBRY Monique, GAREL Gilles
Management of Extreme Situations: From Polar Expeditions to Exploration-Oriented Organizations

MILLOT Michel
Embarrassment of Product Choices 2: Towards a Society of Well-being

N'GOALA Gilles, PEZ-PÉRARD Virginie, PRIM-ALLAZ Isabelle
Augmented Customer Strategy: CRM in the Digital Age

NIKOLOVA Blagovesta
The RRI Challenge: Responsibilization in a State of Tension with Market Regulation
(Innovation and Responsibility Set – Volume 3)

PELLEGRIN-BOUCHER Estelle, ROY Pierre
Innovation in the Cultural and Creative Industries
(Innovation and Technology Set – Volume 8)

PRIOLON Joël
Financial Markets for Commodities

QUINIOU Matthieu
Blockchain: The Advent of Disintermediation

RAVIX Joël-Thomas, DESCHAMPS Marc
Innovation and Industrial Policies
(Innovation between Risk and Reward Set – Volume 5)

ROGER Alain, VINOT Didier
Skills Management: New Applications, New Questions
(Human Resources Management Set – Volume 1)

SAULAIS Pierre, ERMINE Jean-Louis
Knowledge Management in Innovative Companies 1: Understanding and Deploying a KM Plan within a Learning Organization
(Smart Innovation Set – Volume 23)

SERVAJEAN-HILST Romaric
Co-innovation Dynamics: The Management of Client-Supplier Interactions for Open Innovation
(Smart Innovation Set – Volume 20)

SKIADAS Christos H., BOZEMAN James R.
Data Analysis and Applications 1: Clustering and Regression, Modeling-estimating, Forecasting and Data Mining
(Big Data, Artificial Intelligence and Data Analysis Set – Volume 2)
Data Analysis and Applications 2: Utilization of Results in Europe and Other Topics
(Big Data, Artificial Intelligence and Data Analysis Set – Volume 3)

UZUNIDIS Dimitri
Systemic Innovation: Entrepreneurial Strategies and Market Dynamics

VIGEZZI Michel
World Industrialization: Shared Inventions, Competitive Innovations and Social Dynamics
(Smart Innovation Set – Volume 24)

2018

BURKHARDT Kirsten
Private Equity Firms: Their Role in the Formation of Strategic Alliances

CALLENS Stéphane
Creative Globalization
(Smart Innovation Set – Volume 16)

CASADELLA Vanessa
Innovation Systems in Emerging Economies: MINT – Mexico, Indonesia, Nigeria, Turkey
(Smart Innovation Set – Volume 18)

CHOUTEAU Marianne, FOREST Joëlle, NGUYEN Céline
Science, Technology and Innovation Culture
(Innovation in Engineering and Technology Set – Volume 3)

CORLOSQUET-HABART Marine, JANSSEN Jacques
Big Data for Insurance Companies
(Big Data, Artificial Intelligence and Data Analysis Set – Volume 1)

CROS Françoise
Innovation and Society
(Smart Innovation Set – Volume 15)

DEBREF Romain
Environmental Innovation and Ecodesign: Certainties and Controversies
(Smart Innovation Set – Volume 17)

DOMINGUEZ Noémie
SME Internationalization Strategies: Innovation to Conquer New Markets

ERMINE Jean-Louis
Knowledge Management: The Creative Loop
(Innovation and Technology Set – Volume 5)

GILBERT Patrick, BOBADILLA Natalia, GASTALDI Lise,
LE BOULAIRE Martine, LELEBINA Olga
Innovation, Research and Development Management

IBRAHIMI Mohammed
Mergers & Acquisitions: Theory, Strategy, Finance

LEMAÎTRE Denis
Training Engineers for Innovation

LÉVY Aldo, BEN BOUHENI Faten, AMMI Chantal
Financial Management: USGAAP and IFRS Standards
(Innovation and Technology Set – Volume 6)

MILLOT Michel
Embarrassment of Product Choices 1: How to Consume Differently

PANSERA Mario, OWEN Richard
Innovation and Development: The Politics at the Bottom of the Pyramid
(Innovation and Responsibility Set – Volume 2)

RICHEZ Yves
Corporate Talent Detection and Development

SACHETTI Philippe, ZUPPINGER Thibaud
New Technologies and Branding
(Innovation and Technology Set – Volume 4)

SAMIER Henri
Intuition, Creativity, Innovation

TEMPLE Ludovic, COMPAORÉ SAWADOGO Eveline M.F.W.
Innovation Processes in Agro-Ecological Transitions in Developing Countries
(Innovation in Engineering and Technology Set – Volume 2)

UZUNIDIS Dimitri
Collective Innovation Processes: Principles and Practices
(Innovation in Engineering and Technology Set – Volume 4)

VAN HOOREBEKE Delphine
The Management of Living Beings or Emo-management

2017

AÏT-EL-HADJ Smaïl
The Ongoing Technological System
(Smart Innovation Set – Volume 11)

BAUDRY Marc, DUMONT Béatrice
Patents: Prompting or Restricting Innovation?
(Smart Innovation Set – Volume 12)

BÉRARD Céline, TEYSSIER Christine
Risk Management: Lever for SME Development and Stakeholder Value Creation

CHALENÇON Ludivine
Location Strategies and Value Creation of International Mergers and Acquisitions

CHAUVEL Danièle, BORZILLO Stefano
The Innovative Company: An Ill-defined Object
(Innovation between Risk and Reward Set – Volume 1)

CORSI Patrick
Going Past Limits To Growth

D'ANDRIA Aude, GABARRET Inés
Building 21st Century Entrepreneurship
(Innovation and Technology Set – Volume 2)

DAIDJ Nabyla
Cooperation, Coopetition and Innovation
(Innovation and Technology Set – Volume 3)

FERNEZ-WALCH Sandrine
The Multiple Facets of Innovation Project Management
(Innovation between Risk and Reward Set – Volume 4)

FOREST Joëlle
Creative Rationality and Innovation
(Smart Innovation Set – Volume 14)

GUILHON Bernard
Innovation and Production Ecosystems
(Innovation between Risk and Reward Set – Volume 2)

HAMMOUDI Abdelhakim, DAIDJ Nabyla
Game Theory Approach to Managerial Strategies and Value Creation
(Diverse and Global Perspectives on Value Creation Set – Volume 3)

LALLEMENT Rémi
Intellectual Property and Innovation Protection: New Practices and New Policy Issues
(Innovation between Risk and Reward Set – Volume 3)

LAPERCHE Blandine
Enterprise Knowledge Capital
(Smart Innovation Set – Volume 13)

LEBERT Didier, EL YOUNSI Hafida
International Specialization Dynamics
(Smart Innovation Set – Volume 9)

MAESSCHALCK Marc
Reflexive Governance for Research and Innovative Knowledge
(Responsible Research and Innovation Set – Volume 6)

MASSOTTE Pierre
Ethics in Social Networking and Business 1: Theory, Practice and Current Recommendations
Ethics in Social Networking and Business 2: The Future and Changing Paradigms

MASSOTTE Pierre, CORSI Patrick
Smart Decisions in Complex Systems

MEDINA Mercedes, HERRERO Mónica, URGELLÉS Alicia
Current and Emerging Issues in the Audiovisual Industry
(Diverse and Global Perspectives on Value Creation Set – Volume 1)

MICHAUD Thomas
Innovation, Between Science and Science Fiction
(Smart Innovation Set – Volume 10)

PELLÉ Sophie
Business, Innovation and Responsibility
(Responsible Research and Innovation Set – Volume 7)

SAVIGNAC Emmanuelle
The Gamification of Work: The Use of Games in the Workplace

SUGAHARA Satoshi, DAIDJ Nabyla, USHIO Sumitaka
Value Creation in Management Accounting and Strategic Management: An Integrated Approach
(Diverse and Global Perspectives on Value Creation Set –Volume 2)

UZUNIDIS Dimitri, SAULAIS Pierre
Innovation Engines: Entrepreneurs and Enterprises in a Turbulent World
(Innovation in Engineering and Technology Set – Volume 1)

2016

BARBAROUX Pierre, ATTOUR Amel, SCHENK Eric
Knowledge Management and Innovation
(Smart Innovation Set – Volume 6)

BEN BOUHENI Faten, AMMI Chantal, LEVY Aldo
Banking Governance, Performance And Risk-Taking: Conventional Banks Vs Islamic Banks

BOUTILLIER Sophie, CARRÉ Denis, LEVRATTO Nadine
Entrepreneurial Ecosystems (Smart Innovation Set – Volume 2)

BOUTILLIER Sophie, UZUNIDIS Dimitri
The Entrepreneur (Smart Innovation Set – Volume 8)

BOUVARD Patricia, SUZANNE Hervé
Collective Intelligence Development in Business

GALLAUD Delphine, LAPERCHE Blandine
Circular Economy, Industrial Ecology and Short Supply Chains
(Smart Innovation Set – Volume 4)

GUERRIER Claudine
Security and Privacy in the Digital Era
(Innovation and Technology Set – Volume 1)

MEGHOUAR Hicham
Corporate Takeover Targets

MONINO Jean-Louis, SEDKAOUI Soraya
Big Data, Open Data and Data Development
(Smart Innovation Set – Volume 3)

MOREL Laure, LE ROUX Serge
Fab Labs: Innovative User
(Smart Innovation Set – Volume 5)

PICARD Fabienne, TANGUY Corinne
Innovations and Techno-ecological Transition
(Smart Innovation Set – Volume 7)

2015

CASADELLA Vanessa, LIU Zeting, DIMITRI Uzunidis
Innovation Capabilities and Economic Development in Open Economies
(Smart Innovation Set – Volume 1)

CORSI Patrick, MORIN Dominique
Sequencing Apple's DNA

CORSI Patrick, NEAU Erwan
Innovation Capability Maturity Model

FAIVRE-TAVIGNOT Bénédicte
Social Business and Base of the Pyramid

GODÉ Cécile
Team Coordination in Extreme Environments

MAILLARD Pierre
Competitive Quality and Innovation

MASSOTTE Pierre, CORSI Patrick
Operationalizing Sustainability

MASSOTTE Pierre, CORSI Patrick
Sustainability Calling

2014

DUBÉ Jean, LEGROS Diègo
Spatial Econometrics Using Microdata

LESCA Humbert, LESCA Nicolas
Strategic Decisions and Weak Signals

2013

HABART-CORLOSQUET Marine, JANSSEN Jacques, MANCA Raimondo
VaR Methodology for Non-Gaussian Finance

2012

DAL PONT Jean-Pierre
Process Engineering and Industrial Management

MAILLARD Pierre
Competitive Quality Strategies

POMEROL Jean-Charles
Decision-Making and Action

SZYLAR Christian
UCITS Handbook

2011

LESCA Nicolas
Environmental Scanning and Sustainable Development

LESCA Nicolas, LESCA Humbert
Weak Signals for Strategic Intelligence: Anticipation Tool for Managers

MERCIER-LAURENT Eunika
Innovation Ecosystems

2010

SZYLAR Christian
Risk Management under UCITS III/IV

2009

COHEN Corine
Business Intelligence

ZANINETTI Jean-Marc
Sustainable Development in the USA

2008

CORSI Patrick, DULIEU Mike
The Marketing of Technology Intensive Products and Services

DZEVER Sam, JAUSSAUD Jacques, ANDREOSSO Bernadette
Evolving Corporate Structures and Cultures in Asia: Impact of Globalization

2007

AMMI Chantal
Global Consumer Behavior

2006

BOUGHZALA Imed, ERMINE Jean-Louis
Trends in Enterprise Knowledge Management

CORSI Patrick *et al.*
Innovation Engineering: the Power of Intangible Networks

Printed and bound by CPI Group (UK) Ltd, Croydon, CR0 4YY

15/09/2026

14962359-0003